Arado
Ar 234C

David Myhra

Schiffer Military History
Atglen, PA

Acknowledgments

Günter Sengfelder, Nuremberg, Germany—a master fine scale aircraft modeler...none better in the world;

Reinhard Roeser, Lagenhagen, Germany—an uncommon ability to seek out authentic company records, drawings, photographs, and general material due to his encyclopedic knowledge of the Luftwaffe;

Betsy Hertel, Fort Myers, Florida, and Berwick, Pennsylvania—uncommonly supportive wife and best friend.

Mario Marino, Garland, Texas—a gifted digital artist and friend.

Gary Webster, Seattle, Washington—a historical aircraft sleuth who can turn up long-forgotten and mislabeled microfilm with apparent ease;

Dan Johnson, Hampton, Virginia—friend, colleague, and creator of www.luft46.com with its 300,000 annual hits;

Geoff Steele, Arlington, Virginia—friend and colleague whose advice has been incorporated everywhere in this book;

* In captions signifies Arado documents traced and redrawn courtesy of Seweryn Fleischer and Marek Rysi, *Ar 234 "Blitz."*

References

This illustrated history of the AR 234C has benefited from the material presented in the following publications:

Ar 234 Blitz, Monografie 32, Seweryn Fleischer and Marek Rysi, AJ Press, 1997. This thoroughly researched, well written, and richly illustrated book is an outstanding source of information, photographs, and pen and ink drawings of the Ar 234B and Ar 234C;

Arado: History of an Aircraft Factory, Jörg Armin Kranzhoff, Schiffer Military Publications, Atglen, Pennsylvania, 1997. This well-done book provides a complete history of Arado Fluzzeugwerke;

The World's First Jet Bombers, Franz Kober, Schiffer Military Publications, West Chester, Pennsylvania.

Aero Detail 16, *Arado Ar 234 Blitz*, Dai Nippon Kaiga Company, Tokyo, 1993.

Book Design by Ian Robertson.

Copyright © 2000 by David Myhra.
Library of Congress Catalog Number: 00-104359

All rights reserved. No part of this work may be reproduced or used in any forms or by any means – graphic, electronic or mechanical, including photocopying or information storage and retrieval systems – without written permission from the copyright holder.

Printed in China.
ISBN: 0-7643-1182-4

We are interested in hearing from authors with book ideas on related topics.

Published by Schiffer Publishing Ltd.
4880 Lower Valley Road
Atglen, PA 19310
Phone: (610) 593-1777
FAX: (610) 593-2002
E-mail: Schifferbk@aol.com.
Visit our web site at: www.schifferbooks.com
Please write for a free catalog.
This book may be purchased from the publisher.
Please include $3.95 postage.
Try your bookstore first.

In Europe, Schiffer books are distributed by:
Bushwood Books
6 Marksbury Avenue
Kew Gardens
Surrey TW9 4JF
England
Phone: 44 (0) 20 8392-8585
FAX: 44 (0) 20 8392-9876
E-mail: Bushwd@aol.com.
Free postage in the UK. Europe: air mail at cost.
Try your bookstore first.

Arado Ar 234C

The rare *Arado Ar 234C* (C for Caesar), featured in this photo album, was basically the fuselage, wings, and tail assembly of an *Ar 234A/B*. However, it was equipped with four *BMW 003A-1* turbojet engines, instead of the two which had powered the *Ar 234A/B*. The war-time service activities of the *Ar 234B-1* photo reconnaissance and *Ar 234B-2* bomber have been thoroughly dissected by aviation historians, such as *J. Richard Smith & Anthony Kay* in their *German Aircraft of the Second World War* and by *William Green* in his seminal work *Warplanes of the Third Reich*. There is no need to repeat it here. However, the *Ar 234C* was being developed from the *Ar 234B-2*, according to *Arado* officials, with an eye to improving performance and adaptability of their single-seat photo/reconnaissance/bomber. But the question which immediately comes to mind is what additional performance was the *C-* version supposed to provide which the *B-2* version could not do? What need was there to fly faster than the twin *BMW 003A-1* powered *Ar 234B-2's* top speed of 461 mph at 19,685 feet (437 mph with 3,310 pounds of bombs)? Exactly what was the *Ar 234C* to be used for at this late stage of the war? We learn from *Arado* documents that the *Ar 234C's* top speed would be about 530 mph at 19,685 feet altitude...about a 93 mph increase over the typical twin turbojet equipped *Ar 234B-2* bomber version. Likewise, the range of a conventional *Ar 234B-2* was about 683 miles verses 765 miles range for the *Ar 234C*. One can see that this is not a good enough reason for the *Ar 234C*, and given the fact that turbojet engine production was way behind demand, and fuel, difficult to obtain even for the twin-engined *Ar 234B-2*, would be even harder to locate for a four engine version. This book will attempt to discern the real reasons *Arado* engineers added two additional turbojet engines to their basic *Ar 234B-2*, and then will move on to the photographs and drawings of the *Ar 234C*, including remarkably realistic pictures by master scale modeler *Günter Sengfelder* and his *Ar 234C-3* and *Ar 234C AWACS* versions.

The *Ar 234 "Blitz" V1* prototype's first flight occurred on June 15, 1943. Takeoff was accomplished by the use of an *SK 28 204* tricycle dolly which had been developed with the help of *Wittsock Parachute Training Center*. Upon lift-off the dolly usually stayed on the ground, and its forward motion was brought to a stop by its

An Ar 234C-3 *Engländer Bomber* shown out front of its hangar, appearing to be receiving routine maintenance. The *Ar 234 V8*, prototype for the *Ar 234C*, made its maiden flight on 4 February 1944, giving it the distinction as the world's first four-turbojet-powered series aircraft. This author does not consider the experimental *Ju 287 V1* in the running, although its maiden flight was made on 16 August 1944. This is because the *Ju 287 V1* was assembled from bits and pieces of aircraft, both *Luftwaffe* and American, to test the feasibility of a 25° forward swept wing. Scale model, ground equipment, and photographed by *Günter Sengfelder*.

A good view of the Ar 234C-3 with its 4x *Jumo 004B* turbojet engine air intakes. Behind the aluminum cones seen in the center of the air takes are the *Riedel* two cycle, two cylinder starter engines. Just above the nose wheel fork can be seen the built-in holder for its *Lofternrohr* "Lotfe 7D" telescopic bomb sight. Scale model and photographed by *Günter Sengfelder*.

built-in braking parachute. The entire apparatus weighed 1,396 pounds [633 kilograms]. Landing the *Ar 234* was done by a built-in, retractable skid, including a short outrigger-like skid attached to the bottom of each of its two turbojet engines. Eight prototype *Ar 234s* were built, and all eight, which used three-wheel dollies for takeoff, are known as the *Ar 234A* series. The *Ar 234B* series is the improvement of the "A" series, including retractable tricycle undercarriage. Four "B" prototypes were constructed, and the *Ar 234B-0* and *Ar 234B-1* long-range reconnaissance machines were built based on these four prototypes. The *Ar 234B-2* was the high-speed bomber version, with a maximum bomb load of 3,310 pounds [1,500 kilograms]. It was completed in early 1944. A total of 210 *Ar 234B-2* series were built by the time of Germany's surrender on May 8, 1945. On the other hand, only *19 Ar 234C-0* and *C-1* types had been constructed.

A ground-level nose on view of an Ar 234C-3 appearing to be waiting for takeoff clearance. This machine had a very high wing loading of 1,003 pounds per square yard [455 kg/square meters]. Scale model and photographed by *Günter Sengfelder*.

A nose starboard side view of the very clean lines of an Ar 234C-3. These machines had a take-off weight of about 12.3 tons, depending, however, upon armament and bomb load. Scale model and photographed by *Günter Sengfelder*.

Arado Ar 234C

Specifications

	Ar 234B-2	Ar 234C
Type:	Single-seat reconnaissance/bomber	Two-seat bomber/ground attack
Wing span:	45 feet 3 1/2 inches	46 feet 3 1/2 inches
Wing area:	284.167 square feet	284.167 square feet
Length, overall:	41 feet 5 1/2 inches	41 feet 5 1/2 inches
Height, overall:	14 feet 1 1/4 inches	14 feet 1 1/4 inches
Weight, empty:	11,464 pounds	14,400 pounds
Weight, flight ready:	20,700 pounds	24,250 pounds
Weight, maximum:	24,010 pounds	NA
Engine:	2x*Junkers Jumo 004B*	4x*BMW 004A-1*
Speed, Maximum: Sea level -	NA	496 mph at sea level
	437 mph at 19,685 feet	534 mph at 19,685
	373 mph with 3,310 pounds of bombs	NA
Speed, Landing:	99.3 mph	NA
Range, Maximum:	1,013 miles	765 miles
	967 miles with 1,100 pound bomb	NA
	684 miles with 3,310 pound bomb	NA
Climb:	To 19,685 feet with a 1,100 pound bomb...12.8 minutes	NA
	32,810 feet altitude - NA	32,810 feet altitude...16.7 minutes
Ceiling:	32,810 feet attitude	39,370 feet altitude
Bomb load:	3x1,100 pound [*SC 500*] bombs	3x1,100 pound [*SC 500*] bombs
	1x2,205 pound [*SC 1000*] bomb	3xAnti-personnel [*AB 250* or 2x550 pound [*SC 250J*] bombs *AB 500*] bomb clusters
	1x3,086 pound [*PC 1400*] bomb	2x550 pound [*SC 250J*] bombs
	3xAnti-personnel [*AB 250* or 1x*Henschel Hs 294* guided *AB 500*] bomb clusters torpedo missile (air-to-water)	3x*Henschel Hs 293* guided missiles
Cannon:	2x*MG 151* 20 mm with 250 rounds per cannon fixed forward	

A view of an airborne Ar 234C-3 as seen from its port side, above, and behind. It appears to be starting its landing approach because its landing flaps are extended. The *Ar 234C-3* had a design range of 870 miles [1,400 kilometers]. Its range could be increased to 1,180 miles [1,900 kilometers] with two 66 gallon [300 liter] auxiliary fuel tanks under each turbojet engine pod. Scale model and photographed by *Günter Sengfelder*.

A closeup view of the Ar 234C-3 as seen from its port side and featuring its fully extended tricycle landing gear. The nose wheel on the *C-3* had a diameter/width of 770x270 mm, and the two main wheels had a diameter/width of 935x345 mm. On the other hand, the *Ar 234B* had a nose wheel of 630x220 mm and main wheels of 935x345 mm...the same size as the *C-3*. Scale model and photographed by *Günter Sengfelder*.

Arado Ar 234C

The clean smooth lines of an Ar 234C-3 seen from its port side are evidence in this photograph. The *Ar 234C-3's* hinged crew entry hatch is in the open position as seen from its port side. Below the tail fin can be seen the looped wire cable which extends to its braking(drag) parachute located in a box aft in the fuselage. This braking parachute was required to slow the *Ar 234's* landing run. It would be the first aircraft anywhere in the world to have a braking parachute as a standard item. Scale model, ground equipment, and photographed by *Günter Sengfelder*.

Let's go back for a moment and briefly review the development history of the *Ar 234B-1/2* and how they came to be. By the Autumn of 1940, *Hermann Göring's Luftwaffe* had pretty much lost air superiority over England. *Oberstleutnant Theodor Rowehl* was the person in charge of *Luftwaffe* aerial reconnaissance. Using civilian German aircraft belonging to the *Versuchsstelle für Höhenfluge* (Experimental High Altitude Flying Center), *Rowehl* and other pilots compiled aerial photographs of all potential enemy countries before the war. It was *Rowehl* who, as commander of *Auklärungsgruppe Oberbefehlshaber der Luftwaffe (*Reconnaissance Group) after 1939, initially proposed the idea for a single-seat turbojet-powered reconnaissance/bomber. He had been talking to his long-time friend, the chief of *RLM's* planning department *General Ernst Udet*, and together they brought in *Arado's* chief of design *Dip.-Ing. Walter Blume* regarding designs for a turbojet-powered reconnaissance machine. *Arado Flugzeugwerke* was owned by the *Third Reich*. *Walter Blume* was a long time friend of *Udet*;

indeed, their friendship dated back to World War One. Blume had been a fighter pilot in the "Great War." Coming late to the struggle (1917-1918), he was able in only two years prior to the surrender, to score 28 confirmed aerial (kills) victories. Like *Udet*, *Göring*, and the great *von Richthofen*, *Blume* was awarded the *Order Pour le Mérite*, or as it is frequently called, the *Blue Malese Cross* or *Blue Max*. *Theodor Rowehl*, too, had been awarded the *Ritterkruez* for his aerial photo reconnaissance in the 1930s, so here were several highly decorated men planning the next generation aerial photographic flying machines...and these machines would be turbojet-engine powered. All this was at a time when the turbojet engine development was in its infancy, and none of the state-of-the-art engines were even yet available for field use.

In the late 1930s the *RLM* knew that, apart from seaplanes manufactured by *Blohm und Voss Abteilung Flugzeugbau*, *Heinkel AG*, and the *Dornier Werke GmbH*, the German *Luftwaffe* still lacked a good, fast, naval-reconnaissance airplane. Although the *Dornier*

A ground level view of an Ar 234C-3 as seen from its starboard nose side immediately after landing and making its way back to the hangar. Scale model and photographed by *Günter Sengfelder*.

This Ar 234C-3 appears to be factory fresh, that is, without its radio code yet applied, although it is carrying the *Balkenkreuz* on its fuselage and wing surfaces and the *Halkenkreuz* (swastika) on its vertical fin. Scale model, ground equipment, and photographed by *Günter Sengfelder*.

Arado Ar 234C

A Volkswagon staff auto shares the tarmac with an *Ar 234C-3 "Engländer Bomber."* A good view of the aircraft's looped wire cable for its braking parachute can been directly beneath its vertical stabilizer. The cable leads to the parachute storage box on the under side of the fuselage. Scale model, ground equipment, and photographed by *Günter Sengfelder*.

A starboard side view of an Ar 234B-1 (left) and its newer sister, an *Ar 234C-3* (right). The two machines were basically the same aircraft, however, the *234C* had a larger diameter nose wheel and stronger main landing gear. The *Ar 234B-1* is carrying a 66 gallon external tank under its starboard engine. A similar tank would have been carried under its port wing. Scale models and photographed by *Günter Sengfelder*.

A ground level nose port side view of an Ar 234C-3 surrounded by various ground service equipment such the portable 24 volt electrical generating unit/wagon. Scale model, ground support items, and photographed by *Günter Sengfelder*.

A good view of a Ar 234C-3's port side nose. The upside down semi circle shaped "*n*" painted on the port side nose were hand holds for an individual entering the cockpit while the upright semi circle shaped "*U*" was where one placed their foot. Scale model and photographed by *Günter Sengfelder*.

The port side view of a Ar 234C-3 (right) and a *Ar 234B-1* (left). Although they do not appear to be the same, both machines carried an identical-size tail plane assembly. Scale models and photographed by *Günter Sengfelder*.

Two revolutionary machines when they were first introduced to the Luftwaffe about ten years apart: the *Bf 109* (left) and the *Ar 234C-3* (right). Scale models and photographed by *Günter Sengfelder*.

"*Wal*" (whale) flying boat was considered a naval reconnaissance machine, and among the best in its class, the *RLM* viewed it as a temporary solution to their problem. Discussions about a more suitable naval reconnaissance machine took place among the several aircraft manufacturers, and the *RLM* decided by Autumn 1940 that any new reconnaissance airplane would have to be fast—at least 435 mph [700 km/h]—able to cruise comfortably at high altitudes, and have a 1,865 mile [3,000 kilometer] range so that it could fly right around the British Isles. These requirements by *Theodor Rowehl* ruled out all airplanes in the seaplane category. To obtain the high speed, high altitude, and range *Rowehl* wanted, this new machine would have to be powered by the revolutionary new turbojet engines under development in the *Third Reich*. In particular, *Rowehl* wanted *Bayerische Motoren Werke's* (*BMW*) axial-flow *3302* turbojet engine with its 1,500 pounds static thrust at sea level. *BMW* did not design this engine initially. It had been developed by the *Brandenburgische Motorenwerke GmbH* (*Bramo*) of Spandau. *BMW* purchased *Bramo*, and with them their state-of-the-art axial-flow turbojet engine. Afterward, within *RLM* turbojet engine circles the *3302* became known as the *BMW/Brano 003* turbojet engine. Several aircraft manufacturers wanted this early turbojet engine for their proposed aircraft. The *Horten* brothers wanted *BMW/Bramo 3302s* for their all-wing turbojet-powered *Ho 9*. After nearly a year of delay, the *Hortens* were forced by the shortage of time to use the *Jumo 004B*, but it required considerable redesign work to fit the larger turbojet into their all-wing designed to hold the smaller diameter *3302*. *Arado* wanted the *3302* for their new reconnaissance airplane with its promised 1,500 pounds thrust at sea level, espe-

Theodor Rowehl. He and his friend *Ernst Udet* persuaded *Walter Bloom*, in early 1941, of a need for a high-speed and high altitude reconnaissance aircraft for the *RLM's* intelligence-gathering activities. In addition, *Rowehl* required that the new machine be powered by *BMW's* new turbojet engines. Note that *Rowehl* wears the *Ritterkreutz* around his neck. His daring as a high altitude photo reconnaissance pilot was well-known and respected.

cially when *BMW* officials had deemed its turbine ready for field testing in an airframe in the early 1940s. On the other hand, *Junkers Motoren* (*Jumo*) engineers were not yet ready to release their axial flow *Jumo004* turbojet producing 1,980 pounds thrust for field testing. *Ernst Udet*, now a *General* at the *RLM*, was willing to take a chance with the untried gas turbines in a flying machine because these engines appeared, according to their own research done by

The starboard side nose view of a very clean Ar 234C-3 (right) with the starboard side of a *Bf 109* (left). Notice that the *Bf 109* is carrying an external tank. Scale models and photographed by *Günter Sengfelder*.

The Junkers Ju 86R. This was the *Luftwaffe's* unarmed, high altitude photo reconnaissance machine prior to the *Ar 234*, which, among other achievements, had flown over the Scottish highlands. Its ceiling was 42,650 feet [13,000 meters], too high for RAF fighters to reach, at least until late in the war.

Flugbaumeister Helmut Schelp, head of the Special Propulsion Systems branch at the *RLM* since 1939, would provide unheard of rates of climb and straight line speed at altitudes sufficient to make the aircraft immune from enemy *Flak* and fighters.

In late Autumn 1940, *Walter Blume* and his assistant, *Dipl.-Ing. Hans Rebeski*, were invited to submit a design proposal for a turbojet powered reconnaissance airplane capable of 1,865 mile range. In early 1941 the two men discussed their recommendation with *Theodor Rowehl, Ernst Udet*, and his staff at the *RLM*. Initially, *Blume* was not enthusiastic about taking on a new aircraft project...especially one which was to be powered by the new—and untried under field conditions—turbojet engine. *Arado Flugzeugwerke* was fully occupied with projects given them by the *RLM*. Plus, *Blume's* personal favorite project at the time, the twin piston-engined *Ar 240/440 Zestörer (destroyer)*, was in the process of being rejected by the very same people in the *RLM* who now wanted a turbojet-powered long-range reconnaissance machine. *Walter Blume* had not shown much interest in the early turbojet engines and personally doubted whether they were even ready for field operations. Nevertheless, since *Arado Flugzeugwerke* was a Reich-owned aircraft manufacturing company and the Reich could tell them what to design and build, *Blume* and *Rebeski* did what they were told. What the pair from *Arado Flugzeugwerke* were suggesting with their turbojet-engine powered reconnaissance machine, known around *Arado* as *Projekt E.370*, was a single seat monoplane with a shoulder-mounted wing under which were hung two 1,500 pound thrust *BMW/Brano 3302* turbojet engines. The *Jumo 004A's* (later designated the *004B*. The *004A* and *004B* were identi-

Oberst Siegfried Knemeyer. He and *Rowehl* were one time friends and colleagues...both involved in high altitude photo reconnaissance. Note that Knemeyer also wears the *Ritterkreutz*...like *Rowehl*. Among his faults was that *Knemeyer* couldn't stand being #2 in anything. After becoming *Göring's* Technical Advisor in 1943 and chief of the *RLM's* Technical Department (*Udet's* old job), *Knemeyer* saw to it that *Rowehl* was fired.

Left to right: *Ernst Udet* (RLM), *Robert Lucht* (RLM), *Walter Blume* (*Arado*), and *Willy Messerschmitt* (*Messerschmitt AG*). Berlin, 1938.

Walter Blume dressed in his NSFK uniform. He wears his "Ritterkreutz" around his neck, awarded for shooting down 28 enemy aircraft during the 1st World War.

via parachutes. Built-in retractable skids would be used for landing the machine. The *SK* dolly caused considerable problems until it was worked out that it would be released immediately upon lift off and then brought to a stop with the help of a braking parachute. *Udet* and his advisors at the *RLM* appeared to favor the design. But *Udet* committed suicide on November 17, 1941, and his duties were taken over by his successor *Erhard Milch,* a man who was known to favor conventional powered aircraft. Nevertheless, due to the continued influence of *Rowehl, Milch* ordered a wooden mockup of the *E.370* in February 1942. Then, the following April, he ordered *Arado* to construct six prototypes under the *RLM* designation of *Ar 234*. In December 1942, *Milch* ordered 14 more prototypes, thus bringing the total number of prototypes to 20 machines, even though the *BMW 003s* were not yet available for field use. The first test flight by a *234* prototype occurred at the Rheine airfield on the

cal, however, the "*B*" used the absolute minimum amount of scarce metals) had become the engine of choice because *BMW* was having difficulties obtaining the design-rated 1,500 pounds thrust from their *3302* engine due to compressor problems. All *BMW* was obtaining from their *3302* was 570 pounds thrust. They would need to redesign their *003* several times, adding a seventh compressor stage, and so on, to obtain the design-rated thrust. They would not achieve it until the end of 1942 with an engine they were calling the *003A-0.*

With the *Ar E.370, Blume* and *Rebeski's* design called for a fuselage which had a stressed-skin semi-monocoque with so-called "top hat" section longerons and "Z" section formers and stringers. The two-spar stressed-skin wing rested on a reinforced box-girder section of the fuselage and bolted to the upper four longerons at four places. The pilot's cockpit, which was extensively plexiglassed and pressurized by air diverted from the *Jumo 004A's* compressors, occupied the entire nose of the machine. Aft the cockpit were the fuel tanks. The wing featured a dual taper on the leading edge, its trailing edge carrying Frise ailerons of very narrow width and hydraulically operated hinged, two-section flaps. The *Jumo 004As* were hung from beneath the wing with three attachment bolts...two in the front, and one at the rear of the turbojet engine. Each turbojet engine was covered by sheet metal cowling. The main gear storage proved a difficulty, since the shoulder-mounted wing did not provide sufficient space to permit the wheels and ole struts to be retracted up and stored in the wing. While *Arado's* designers worked on a solution, in the interim, the *RLM* chose to allow *Arado* to utilize the *SK 28 204* three-wheel, steerable dolly with hydraulic brakes on the two rear wheels. The idea was that the *SK* dolly would be jettisoned at about 300 feet after lift off and returned to the ground

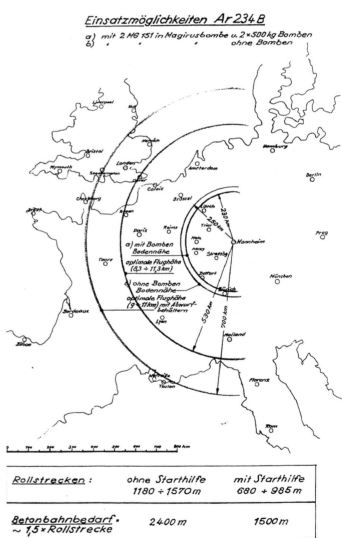

This illustration shows the range which *Rowehl* was seeking with a turbojet-powered *Ar 234*. With a machine taking off from the German city of Mannheim, for example, *Rowehl* wanted his photo reconnaissance pilots to be able to reach as far west as the major port city of South Hampton, as well as beyond London itself.

Arado Ar 234C

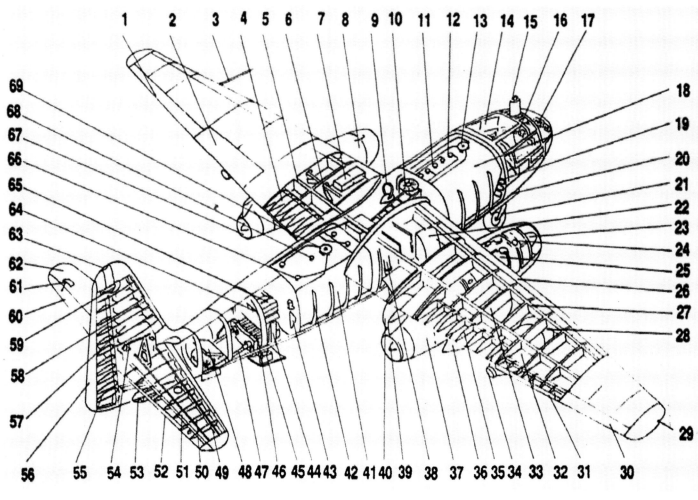

01 - port side aileron
02 - port side inboard landing flap
03 - pitot tube
04 - port side outboard landing flap
05 - port side wing's secondary/rear spar
06 - port side flap actuator
07 - *Funkgerät FuG 25a* IFF (radio instrument)
08 - port side *Jumo 004B* engine nacelle
09 - *Funkgerät FuG 25a* (radio instrument antenna)
10 - radio-direction-finder loop antenna
11 - forward/main fuel tank fuel gauge sensor
12 - forward/main fuel tank fuel pump
13 - forward/main fuel tank filler opening
14 - cockpit entry/egress hatch
15 - *BZA* periscope
16 - steering yoke
17 - instrument panel
18 - forward fuel tank
19 - 3 oxygen tanks
20 - nosewheel cover doors
21 - nosewheel assembly
22 - starboard side main wheel cover door
23 - *Riedel* turbine starter motor...inside aluminum cover cone
24 - oil tank
25 - starboard *Jumo 004B* turbojet engine

26 - starboard wing's leading edge
27 - starboard side wing's main/forward spar
28 - *HWK 500A* bi-fuel liquid rocket take off booster parachute holding box
29 - starboard side wing tip (green) navigation light
30 - starboard side wing aileron
31 - starboard side wing aileron trim tab
32 - starboard *HWK 500A* bi-fuel liquid rocket take off booster nacelle
33 - starboard wing internal ribs
34 - nozzle of the starboard *HWK 500A* bi-fuel liquid rocket take off booster
35 - starboard wing internal trailing ribs
36 - external 600 liter fuel tank
37 - starboard *Jumo 004B* engine rear nacelle
38 - starboard main wheel oleo leg
39 - starboard *Jumo 004B* tail cone
40 - starboard main wheel
41 - aft fuel tank
42 - forward fuselage ribs
43 - tail plane control push rods
44 - starboard side fixed rearward firing *MG 151* 20 mm cannon ammunition feeder
45 - starboard side *MG 151* 20 mm cannon shell case exit port

46 - starboard side *MG 151* 20 mm cannon
47 - starboard side *MG 151* 20 mm cannon barrel
48 - compartment for braking parachute
49 - starboard horizontal stabilizer's main spar
50 - starboard horizontal stabilizer's internal frames
51 - starboard horizontal stabilizer's auxiliary spar
52 - starboard horizontal trim tab
53 - braking parachute braking cable
54 - vertical stabilizer's (rudder) control push rods
55 - vertical stabilizer's (rudder) internal ribs
56 - vertical stabilizer's (rudder) trim tab
57 - vertical stabilizer's internal ribs
58 - vertical stabilizer's internal leading edge post
59 - vertical stabilizer's (rudder) actuator lever
60 - port horizontal stabilizer elevator
61 - port horizontal elevator actuator lever
62 - port horizontal stabilizer
63 - vertical stabilizer's (white) navigation light
64 - rear fuselage frame
65 - port side fixed rearward firing *MG 151* 20 mm cannon
66 - aft fuel tank pump
67 - port side *Jumo 004B* turbojet engine nozzle
68 - aft fuel tank fuel gauge sensor
69 - aft fuel tank fuel filler port

Arado Ar 234C

A fully outfitted Ar 234B-1 reconnaissance machine as seen from it nose port side. Scale model and photographed by *Dan Johnson*.

evening of July 30, 1943, and was powered by *Jumo 004A* turbojet engines. Prototypes *234V1* through *V7* all had to be powered by *Jumo 004s*. The Rheine airfield was located north of Münster, in Westphalia, and the *Ar 234 V1's (TG+KB)* test pilot on its historic first flight was *Arado's* own company test pilot, *Flugkapitän Heinz Selle*. *Selle* would lose his life on October 2, 1943, in the crash of the 2nd *Ar 234* prototype. An engine fire in the port *Jumo 004* at 29,363 feet [8,950 meters] caused the machine to go into a dive because the aileron linkage and electrical wiring had been burned through. On November 5, 1943, *Erhard Milch* instructed *Arado* to make plans to construct 100 *Ar 234Bs*. *Milch* would later increase the order by another 100 in December, in addition to requiring the entire order of 200 to be fulfilled by the end of 1944.

On November 26, 1943, the *Ar 234V3*, *werk nummer 130003*, was demonstrated in front of *Hitler* in Insterburg. *Hitler* is reported to have been highly enthusiastic, and immediately called for the production of the 200 machines...all to be built as high-speed bombers, or "Blitz bombers." Hitler saw these machines as an ideal instrument for striking against a potential Allied advance. And fast it

Ground level starboard side view of an Ar 234B-1 reconnaissance machine. Scale model and photographed by *Dan Johnson*.

was. *Horst Götz*, piloting the *Ar 234V5*, reached 559 mph [900 km/h] in a shallow glide from 4,971 feet [8,000 meters], and he felt that it could have achieved even higher speeds were it not for the tail assembly's controlling surfaces experiencing shuddering from the uneven air flow in the narrow gap between the fuselage and turbojet engines.

The Ar 234B-1 reconnaissance machine as seen from behind. Scale model and photographed by *Dan Johnson*.

Arado Ar 234C

A nose on view of the Ar 234B-1 reconnaissance machine as seen from above. Scale model and photographed by *Dan Johnson*.

A ground level nose on view of the Ar 234B-1 reconnaissance machine. Scale model and photographed by *Dan Johnson*.

In late 1943, an assembly line for the Ar *234B-1* series prototypes, all powered with *Jumo 004B* turbojet engines producing 1,980 pounds thrust each, was set up at Alt Lönnewitz in what is now the Czechoslovak/German border. This was pretty much in parallel with the assembly line for the pre-production *Ar 234B-O* reconnaissance aircraft. During the initial flight testing the *Ar 234's Jumo 004B* turbojet engines operated on *K1* diesel oil, however, shortages forced the test program to use the more readily available *J2* lite oil which had about the same heat value. The *Jumo 004B* was self-started by a *Riedel* two-cylinder gasoline starter motor geared to the compressor shaft and mounted in the *004's* air intake. The *Riedel* would run the *004* turbine up to its self operating level of 3,000 revolutions per minute (rpm). At 3,000 rpm, gasoline was injected into its combustion chambers and ignited electrically. At 6,000 rpm, the pilot switched to *J2*, replacing gasoline for the remainder of the flight.

Operational use of the *Ar 234B-1* began almost as soon as prototypes and pre-production machines were rolling off the production lines at Alt Lönnewitz. The first *Sonderkommando*, Sd.Kdo.

Götz, was established at Rheine with four *Ar 234B-1s* flying on reconnaissance flights over the British East Coast harbors between the Thames Estuary and Yarmouth. In addition, Sd.Kdo.*Götz* went flying over South England to learn if the Allies were making preparations for an invasion of the Netherlands. In its photo-picture-taking role, the *Ar 234B-1s* were equipped with two *Rb 50/30* or *Rb 75/30* cameras and known as *Ar 234B-2/b*. On the other hand, several photo reconnaissance versions were equipped with a single *Rb 75/30* and a single *Rb 20/30* camera. Photo reconnaissance usually took place at about 29,530 feet [9,000 meters] altitude. Each camera's magazine contained about 395 feet [120 meters] of film. The interval between exposures was 10 to 12 seconds, and there was usually a 60% longitudinal overlap.

Several *Ar 234B-1s* were equipped to carry up to 3,300 pounds of bombs and were called *Ar 234B-2*. This usually included one 1,100 pound [*SC 500*] bomb hung beneath the fuselage, plus two more, one beneath each *Jumo 004B* engine nacelle, and all on *ETC 504* racks. Or, the *234B-2* could carry a single 3,086 pound [*SC*

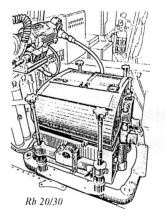 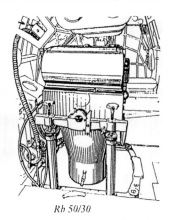

A pen and ink drawing of two other models of aerial reconnaissance cameras: (left) a **Rb 20/30** and (right) a **Rb 50/30**.

This Luftwaffe ground crewman has his arms wrapped around an *Rb 50/30* reconnaissance camera which has been removed from an *Me 262A* reconnaissance machine.

29,000 to 30,000 feet, but all attempts were unsuccessful in getting high enough and close enough to accurately shoot at them, hoping to force one down over Allied occupied territory. The Allies were not able to capture an intact *Ar 234B* until February 24, 1945, when an *Ar 234B-2* coded *F1+MT* from *9./KG 76* suffered a flame out in one of its *Jumo 004* turbojet engines and was forced down by U.S. *P-47* "*Thunderbolts*" outside the village of Segelsdorf. The following day the U.S. 7th Army occupied Segelsdorf, and the *Ar 234B-2*, which had belly-landed in a farm field, was captured intact, although heavily damaged. The machine was disassembled by British Intelligence and removed to RAF-Farnborough for repair, flight testing, and evaluation. A total of 210 *Ar 234Bs* had been completed at Alt Lönnewitz when production ceased due to its take over by the Soviet Red Army. It is believed that 19 *Ar 234Cs* were completed at Alt Lönnewitz, too, with dozens more in various stages of completion and all captured by the Red Army.

Two Plus Two Equals Four:
The first four turbojet powered bombing machine.

But why go to four turbines instead of two? In conversations this author had post-war with former *Arado* chief aerodynamicist

A Luftwaffe ground crewman has removed the *Rb 50/30* reconnaissance camera from an *Ar 234B-1*.

1000] bomb hung beneath its fuselage. Armament usually included a pair of rear-ward firing *MG 151* 20 mm cannon. In aiming the *MG 151* rear-ward, a *PV1B* sighting head from a *RF2C* periscope was normally fitted for the pilot. The standard *Ar 234B-2* bomb-carrying machine was known as the *Ar 234B-2/p*, *-2/r*, or *-2/pr* depending on the equipment installed, such as "*P*" if the *Patin PDS* three-axis autopilot with an *LKS 7D-15* overriding control equipped to be used in a "pathfinder" role had been installed. Most *Ar 234Bs* carried the *Patin PDS* autopilot. If external fuel tanks were installed then the machines became known as the *Ar 234B-2/bpr*. The only protective armament for the pilot was a 15.5 mm steel plate attached to the bulkhead behind the pilot's head, providing a small measure of protection to his head and shoulders.

In conversations between *Horst Götz* and this author, *Götz* described how *Sd.Kdo.Götz* had been formed with men coming from the *Versuchsverband Ob.d.L.* According to *Horst Götz*, many successful reconnaissance missions were flown. He and his pilots relished with delight as Allied (British) stripped down fighters sought to reach and shoot down the high-flying *Ar 234B-1s* at altitudes of

Horst Götz...leader of the *Ar 234B* reconnaissance group *Sd.Kdo.Götz*. Late 1944.

Arado Ar 234C

Horst Götz's wrecked *Ar 234B* with its tail assembly sheared off by a *Fw 190* and a pilot not looking where he was going.

Rüdiger Kosin, he said that the *RLM's* interest in producing bombers increased with the successful flights of the Me 262, and even more so with the *Ar 234B-2. Kosin* recalled that in late 1942 both *Hitler* and *Göring* were insisting that bombers were needed to carry on the air war to England. It was *Hitler's* mania for retaliation, rather than defense, that the *Ar 234B-2* was now being thought of as a potential long-range bomber...powered by four turbojet engines instead of two. *Kosin* said:

> We had to do the modifications or risk being sent off to the Russian Front. The *Führer* could think only of retaliation and attack, rather than defending the German population, which was suffering hundreds of thousands of casualties from Allied bombers. You see, *Göring,* who had once promised that no enemy bombers would ever fly over the *German Reich,* was

Horst Götz (right) with author (left). Nieby, West Germany. 1984.

now feeling the full fury of *Hitler's* rage, seeing that they were roaming pretty freely over German soil. It was *Göring,* then, acting as his *Führer* wished...rather than taking increased measures for defense. *Göring* appointed an *"Angriffsführer Engländ"* (attack leader - England) to carry out reprisal bombings. Then again, any small raids with the twin turbojet powered *Ar 234B* would likely accomplish little, especially against a British air defense that had been massively strengthened since Germany's losing battle for Britain four years prior. So a four motored *Ar 234* known as the *"C"* would be built to keep the *Führer* happy, and would be known as the *Ar 234C,* or *"Die Engländer Bomber."* *Walter Blume* did not want to be involved in the *Ar 234C,* therefore, he gave me the responsibilities, and so I became the project's director.

In fact, on November 11, 1943, *Hitler* ordered that daily reports be submitted to him on the production of the bombers *Me 262* and *Ar 234B.* After all, hadn't *Hitler* fired *General Erhard Milch* on May 23, 1944, for subverting series production of the *Me 262* into a fighter and away from *Hitler's* demand that it be a *Blitzbomber*?

> "For years I have demanded from the Luftwaffe a 'speed bomber,' which can reach its target in spite of enemy fighter defense. In the aircraft you present to me as a fighter plane I see the 'blitzbomber' with which I will repel the invasion in its first and weakest phase. Regardless of the enemy's air umbrella, it will strike the recently landed mass of material and

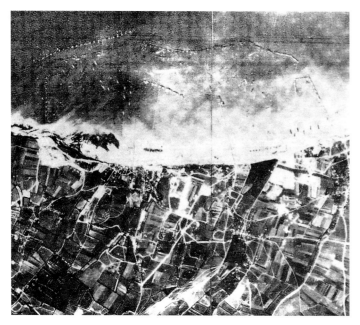

Normandy beaches during the Allied invasion as photographed from a Ar *234B*.

troops, creating panic, death, and destruction. This is the 'bliztbomber'...of course, none of you thought of that."

On May 23rd, 24th, 25th, and 29th, 1944, a series of crucial secret conferences on the *Luftwaffe's* emergency aircraft construction program took place at the Obersalzburg, with *Hitler* presiding on the first day. When the subject of the *Me 262* came up, *Hitler* turned to *Milch*, asking him how many of the finished aircraft were able to carry bombs? *Milch*, who apparently overlooked *Hitler's* sole interest in the *Me 262* as a *Blitzbomber*, replied:

> "None, my *Führer*; the *Me 262* is being built exclusively as a fighter aircraft."

Hitler burst into one of his fire-belching rages, accusing *Göring*, *Milch*, and the entire *Luftwaffe* of unfaithfulness, disobedience, and unreliability. *Milch* was dismissed as head of *Luftwaffe* armaments that very day. The day Milch was fired, *Göring* issued his famous order in which he declared that the *Führer* had ordered the *Me 262*

The Me 262V5 with its fixed landing gear while it was being tested with 2x*Borsig Ri 502* rocket take off assist solid fuel rockets with a 12 minute burn time. Shown here without its *Ri 502* rockets attached beneath the fuselage.

to be built exclusively as a *Schnellstbomber* (super-fast bomber). In case there was anyone in Germany who still did not understand, *Hitler* himself issued a short, curt "*Führer Order,*" too, stating:

> "The *Führer* orders that the *Me 262* and *Ar 234* are to be, in fact, bombers, and not fighters."

Furthermore, during this secret conference at the Obersalzburg on 23 May 1943, *Hitler* held a meeting with German aircraft builders, from which *Göring* and *Luftwaffe* leaders were barred. *Hitler* demanded a bomber capable of attacking London by day and night from altitudes above the effective reach of interceptors, and also able to attack Allied convoys far out in the Atlantic. When *Ernst Heinkel* mentioned that a conventional version of his *He 177* could meet the *Führer's* demands...right now, *Hilter* reminded *Heinkel* of the *He 177's* troubles with its coupled *Daimler-Benz 610* piston engines. *Hitler* told him that to design the heavy machine so it could dive bomb, too, was idiocy. *Heinkel* said no more. However, shortly after the May 23rd conference, the *RLM* gave *Heinkel* permission to construct the *He 177* with four separate engines, and it became known as the *He 277*. All existing older model *He 177s* with their coupled *Daimler-Benz 610* engines were ordered scrapped. The order appears to have been ignored, but beginning in January 1944, operations against English were attempted with the coupled *Daimler-Benz 610* engined *He 177* bombers, as the *Luftwaffe* did its best to carry out *Hilter's* demands for a retaliatory air offensive. It was shown that British night fighters and anti-aircraft were ineffective against the fast, high-flying *He 177s*, but its own mechanical failures caused losses that were unreasonably high compared with the small effect of the night bombing for all the great efforts expended.

By the beginning of March 1944, after the *He 177* had obviously been a failure in attacks on Britain, *Hitler* declared angrily in a *Führer Conference*:

Adolph Hitler was determined to obtain a "*Engländer Bomber.*" That would be a machine which could reach altitudes and speeds unmatched by British "*Spitfires,*" drop its extensive bomb load, and return to its base in Germany. No air machine seemed to be able to fill that role, not even the *Me 262*, until *Arado* presented *Hitler* with their *Ar 234C* prototype "*Die Engländer Bomber.*"

Arado Ar 234C

Me 262 with *2x250* kilogram [551 pounds] bombs.

An Ar 234B without any external items, such as bombs or fuel tanks.

This junk machine (the *He 177*) is naturally the greatest shit that has probably ever been manufactured. It is the flying *Panther*, and the *Panther* is the creeping *Heinkel*! (the Panther was a German tank which was the source of much trouble in its development stage.)

The *He 177* was a real catastrophe for the *Luftwaffe*. More than 50 valuable crews were lost in its testing, while the 1,146 aircraft which were produced represented virtually a total waste. About 5,750 badly needed fighters could have been built with the same expenditure of materials and effort. So, here is the reason for the *Ar 234C* and why the twin turbojet *Ar 234B-1* reconnaissance machine was being redesigned with four turbojet engines so that it could carry bombs, if not to America then certainly to Great Britain. This *Arado* four engine short range bomber version (1,000 miles) was being called the *Ar 234C "Snellistbomber."* Maybe it should have been titled: *"Die Engländer Bomber,"* because this is what its primary duties would have been.

The *Ar 234C...Hitler's "Die Engländer Bomber"*
Arado had been given the *BMW 003A-1* turbojet engine producing 1,760 pounds [800 kilograms] thrust at sea level for its *Die Engländer Bomber* experiment. Before installing four turbojet engines on the formerly twin turbojet engine *Ar 234B-1*, *Rüdiger Kosin* and his *Arado* engineers wanted to know how the four turbojet engines should be mounted. All four would be hang under the wing just as they had been on the *Ar 234B-1*...but should the four engines appear as two paired engines in one cowling per wing, or two individual and widely spaced turbines per wing? In order to arrive at a definite answer, two prototypes were to be built using the wing and fuselage from a standard *Ar 234B-1*. The air frame with four individual widely spaced *BMW 3302* engines would be known as the

The Ar 234 V9 shown in this photograph with 3x*SC 500* kilogram [1,102 pounds each] bombs. A similar arrangement would have been used for the bomb-carrying variations of the *Ar 234C* series.

The Ar 234 V9. Notice that this machine has had a section of its lower center fuselage removed and replaced with a concaved section to help accommodate a SC *500* bomb with less drag.

Arado Ar 234C

Port side rear view of the Ar 234 V9 carrying 3xSC 500 bombs. This arrangement would have been similar to the bomb-carrying version of the Ar 234C.

A close up of a Ar 234B carrying a single SD 500 bomb under its fuselage...the same way a Ar 234C bomber version would have done it.

Ar 234 V6 coded *GK+IW*. The air frame with two *BMW 3302* engines paired in a single cowling per wing would be known as the *Ar 234 V8* coded *GK+IY*. Modifications on both *Ar 234B-1* airframes began at about the same time, however, the *Ar 234 V8* was finished first and began its flight trials on February 1, 1944. The second prototype, with four individual widely spaced engines and known as the *Ar 234 V6*, was finished about two months later...April 8, 1944.

Attaching two additional engines on the *V6* was relatively simple. *Rüdiger Kosin* and his staff of *Arado* engineers attached a third and fourth engine to the same mounting points that the *Ar 234B-1* had used for its *HWK 500* takeoff rocket boosters. On the *V8*, with its two paired engines, *Arado* engineers attached two *BMW 3302s* to a single frame and mounted it to the attachment points formerly carrying only one engine. After testing both versions thoroughly, *Kosin* told this author that they settled on the paired engine arrangement as the more effective, efficient, and practical way to hang multiple engines on their *Ar 234Cs* aircrafts' wings. The individual unit arrangement, *Kosin* found, gave the result of reducing yaw stability, while at the same time increasing the machine's roll moment during a sideslip. These two conditions could result in the aircraft experiencing tumbling at high altitudes. With the *V6* and its paired engines, *Kosin's* engineers discovered that the paired arrangement placed the engines too close to the fuselage, creating a pressure build up on the tail control surfaces due to regions of accelerated air flow caused by the narrow gap between the engine cowling and the fuselage.

Rüdiger Kosin obtained a third *Ar 234B-1* and also fitted it with 4x*BMW 3302* engines in a paired configuration and producing 1,760 pounds thrust each at sea level. This 3rd test version was known as the *Ar 234 V13*, and it was being flight tested in late August 1944. So, with the 2nd airframe with paired engines now built, *Ko-*

Hilter was being led to believe the new *Messerschmitt Me 262* was being fitted out as a bomber, at least as a short-range bomber. When he asked *General Erhard Milch* about it at a meeting held at Obersalzburg 23 May 1944, how many *Me 262* bombers had been built, *Milch* replied "none, my *Führer*, they are all being built as fighters." *Hilter* fired him later that day.

The Me 262 fitted out as a fighter and not as a *"blitz bomber,"* which *Hilter* was expecting. *Milch* would lose his job in the *RLM* as a result. He was lucky not to have lost his life over this matter.

SC 2500 "Max" [5,512 pounds] was the *Luftwaffe's* heaviest free-fall bomb.

sin had pretty much decided that the paired engine arrangement would be the layout for all their *Ar 234C* series designs. He told this author that each day brought new duties for the proposed four-engined *Ar 234*: bomber versions, reconnaissance versions, nightfighter versions, even day fighter versions. It came to be bewildering, with 2 engine versions, 4 engine versions, and even a *HWK 509C* powered high altitude reconnaissance version wanted by the *RLM*. He told this author that about the time *Walter Blume* told him to redesign the *Ar 234* with four engines, he was sitting in one of the windowless rooms at the *RLM*-Berlin in Autumn 1944; windowless because Allied bombing had knocked them all out.

I looked about inside the *RLM* room where I was and then outside to the great city of Berlin. Now, the *RLM* was asking for all these four engined-powered *Ar 234* variations. I still recall that all this planning was simply stupid, because we had lost the war way back with our failure in the Battle of Britain. Now I looked out the window and saw Berlin suffering de-

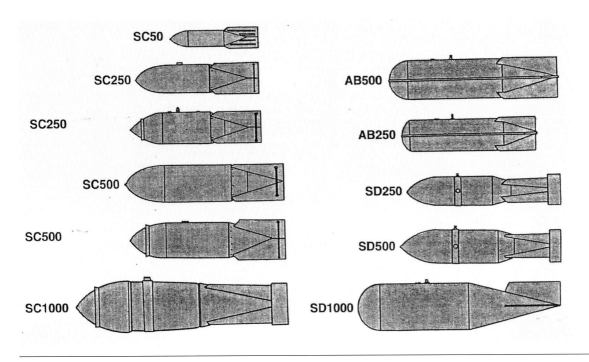

Ten of the Luftwaffe's most commonly used bombs...from the *SC 50* [110 pounds] to the *SD 1000* [2,205 pounds].

Arado Ar 234C

This SC 2500 "Max," from about 1942, had an explosive charge weighing 3,748 pounds [1,700 kilograms].

struction day after day...in this environment we were supposed to design a heavy four-engined aircraft in dozens of variations? Impossible.

In addition, to be sure, *Kosin* took a 4th basic *Ar 234B-1* airframe which had been powered by 2x*Jumo 004Bs* producing 1,980 pounds thrust at sea level and exchanged them with 4x*BMW 003A-1s*. They called this re-engined machine the *Ar 234 V15*. *Arado* felt that they had to make this switch because they were having considerable difficulties in regulating fuel flow during and after acceleration during the test flights with the *BMW 003A-1s*. Perhaps *BMW* engineers didn't experience thrust flow on bench testing back at their Berlin laboratories, but out in the field under actual conditions it acted much differently.

By March 1944 definite plans for producing the *Ar 234C* into two sub-types were established by the *RLM*. The sub-types included the *C-1* - reconnaissance version, and the *C-2*- bomber version. In

A AB 1000 "*Abourfbehälter*" bomb which was filled with small fragmentation bombs.

The twin piston engined late 1930s designed Heinkel He 111 flew to slow and low to be an "*Engländer Bomber*." It had been routinely shot out of the skies during Germany's battle for Britain.

Arado Ar 234C

The He 177's four motors coupled to form two motors and turning two large four-bladed propellers had been sent to England in late 1944. But its all around effect from the bombing was small due to insufficient numbers, and losses caused by its own mechanical problems. But the *He 177* wasn't an easy target even for the British *"Tempest,"* perhaps the highest flying piston-powered fighter in the world at that time.

September 1944, the *RLM* announced two more sub-types. These two sub-types included the *C-3* - a bomber variant, and the *C-4* - a reconnaissance variant. Both the *C-3* and *C-4* version had a higher cockpit cabin roof and a fully pressurized cockpit. Additional variations included: a bomber *C-5*; *C-6*, a long-range reconnaissance machine; *C-7*, a night-fighter; and the *C-8*, projected as a bomber. Prototypes *V19* (coded *PI ι WX*) through *V30* may have been constructed. However, due to the lateness of the hour, none were thoroughly evaluated for their intended roles. In some reports it is mentioned that perhaps only five prototypes were built...*V21* through *V25*. The record is not very clear. The same is true for the *V19*'s maiden flight at Sagan. One report states that the *V19*'s flight, which was considered to be the prototype for the entire C-series, occurred on September 30th, while another report claims that it took place on October 16, 1944. Nevertheless, it was also fitted with experimental air brakes and a larger nose wheel fork and tire. Prototype *V20* appears to have been similar to the *V19*, however, for the first time its cockpit cabin was pressurized. For all practical purposes, the *V20* was supposed to have contained all the refinements and modifications *Kosin's* engineers wanted in the *Ar 234C* series: a redesigned cockpit cabin, which essentially had a higher roof to improve all around visibility; a rearward firing 2x*MG 151 20* mm

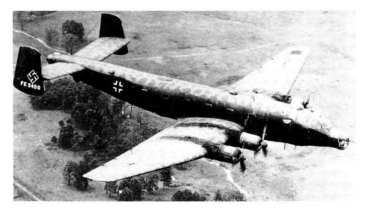

The four piston motored Junkers Ju 290 could have been *Hilter's "Engländer Bomber,"* but it was also too slow. This is a post-war photograph of the *Ju 290*, because the machine has a U.S. foreign equipment *"FE"* code painted on its starboard rudder beneath its *Halkenkruez (swastika)*.

cannon, plus 2x*MG 151 20* mm cannon fitted on the fuselage beneath the cockpit; and 3x*ETC 504* bomb racks. Prototype *V21* is reported to have made its first flight on November 5th, and *V22* on December 21, 1944, all from Sagan (now known as Zagán, Poland). How well did the world's first four turbojet powered ma-

The Messersch*mitt Me 264* with its dual rudders, landing gear full down, and shown during a flight test. A rare sight. The four piston engined *Me 264* prototype design was also too slow to be a true *"Engländer Bomber."* Initially offered by *Messerschmitt* as an *"Amerika Bomber"* it lacked the range needed for a round trip flight to America.

The massive Blohm & Voss six piston engined *Bv 238* sea plane may have had the range to fly right around the British Isles with a full load of bombs, however, it would have been easily shot down due to its slow speed.

chine perform? *Rüdiger Kosin* said *Arado* test pilots reported that their four-motored *Ar 234C* performed pretty much like the twin-motored *Ar 234B-1*.

In *Arado* documents obtained post-war, mention is made of a machine known as the *Ar 234 V16*. It appears that it was to have been powered by 2x*BMW 003Rs*...the combined turbine and bi-fuel liquid rocket engine *BMW* engineer *Count Helmuth von Zborowski* was seeking to perfect. He had been with *Eugen Sänger* for years as a research assistant on the *Sänger* orbital bomber project. Zborowski's *BMW 003R* was not ready during the war for anything other than experimental testing. If so, it would have had a combined thrust of 4,510 pounds thrust at sea level. In addition, the V16, which was known also as the *Versuchsflügel #1*, was to have had the new crescent wing developed by *Rüdiger Kosin*. The V16's thick crescent wing was constructed out of wood. It is also reported that the *Ar 234 V18*, known also as the *Versuchsflügel #2*, had been fitted with a thin crescent wing, not of wood, but constructed out of aluminum. It is not known to this author if the V16 had actually been test flown with its 2x*BMW 003R* engines installed. *Arado's* engineers were anxiously testing additional wing profiles for their

The six piston engined Focke-Wulf Fw 390 was also too slow to be considered an effective *"Engländer Bomber."*

An FZG 76 coming down over London in the Piccadilly Station area. A frightful sight.

The Fieseler Fi 103 (also known as the *Flakzielgerät FZG 76, V1*, and "*Doodlebug*" to the people in England) was *Hitler's* next attempt to rain bombs down on England in the absence of a true "*Engländer Bomber*." Ground-launched *FZG 76s* from sites in northern France began on 13 June 1944, about six days after the Allies had made a beach landing on the western coast of France. Ground launching of *FZG 76s* from France was pretty much over by early August, lasting about ten weeks, because the Allies had captured and/or bombed all their launch sites. By the time it was over, 9,017 "*Doodlebugs*" had been launched toward England from France. British records list 6,725 had been reported seen over England.

234C machines. The *Ar 234 V26* was also known as the *Versuchsflügel #3*, and had been constructed to test *Arado's* new all metal laminar profile wing. *Versuchsflügel #4* was known as the *Ar 234 V30*. It had, or was to have had, swept-back all metal laminar flow wings. There are reports from *Arado* that four *Ar 234C* prototypes, the *V18, V21, V26*, and the *V30*, had in fact been constructed, and that they had been destroyed by *Arado* personnel to prevent their secret wing profiles from going to the enemy...the Soviets.

The only major difference between the *Ar 234B-1* and the *Ar 234C*, other than the number of turbojet units, was the redesigned cockpit in the *Ar 234C*. It was designed to be pressurized, and this meant that instead of having only a single sheet of plexiglass as was found on the *Ar 234B-1*, the *Ar 234C* was to have two sheets of laminated plexiglass throughout its cockpit. The first *Ar 234C* series with cabin pressurization began with the *Ar 234 V20*. However, an experimental cockpit cabin pressurized prototype *Ar 234*

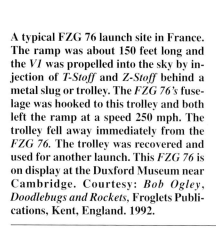

A typical FZG 76 launch site in France. The ramp was about 150 feet long and the *V1* was propelled into the sky by injection of *T-Stoff* and *Z-Stoff* behind a metal slug or trolley. The *FZG 76's* fuselage was hooked to this trolley and both left the ramp at a speed 250 mph. The trolley fell away immediately from the *FZG 76*. The trolley was recovered and used for another launch. This *FZG 76* is on display at the Duxford Museum near Cambridge. Courtesy: *Bob Ogley, Doodlebugs and Rockets*, Froglets Publications, Kent, England. 1992.

Arado Ar 234C

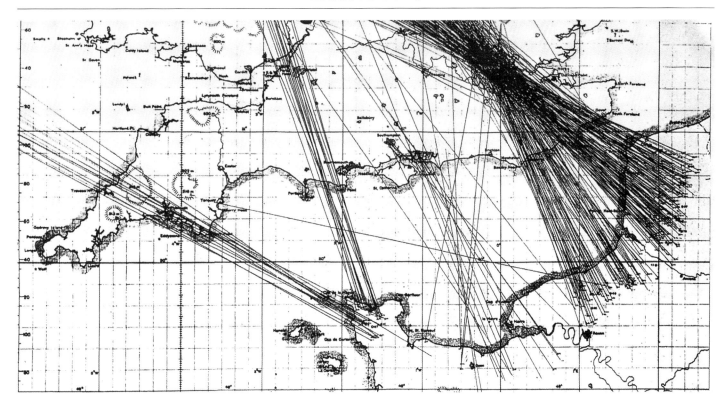

A captured German graph showing the proposed FZG 76 launch/firing lines to southern England from northern occupied France. About 5,500 people lost their lives to the *FZG 76* and its 1,874 pound [850 kilogram] warhead/bomb.

with four turbojet engines, the *Ar 234 V13*, is reported to have reached an altitude of 42,000 feet.

The *Ar 234C* series had several minor differences between it and the *Ar 234B-1*. These included some aluminum reshaping, changes in aileron design, and a larger nose wheel fork and wheel. It is not known for sure how many initial production *Ar 234C-Os* and *C-1*s were constructed before the *RLM* chose to construct only the basic *Ar 234C* in a multi-role version and known as the *Ar 234C-3*.

London.

One of the most famous photos published in England during the FZG 76 blitz. This photo appeared in the *Daily Mirror* and shows an eleven-year old girl being carried in the arms of a fireman. Her home and others in Leytonstone have just been hit by an *FZG 76*.

London, 3 August 1944. A FZG 76 falls behind the Law Courts, seen left in the photograph.

Ar 234 Versuchsmuster (A, B, and C prototypes)

V1 - Ar 234A version prototype. 2x*Jumo 004A* turbojet engines, *werk nummer 130001*, coded *TG+KB*, first flight 15 July 1943, crashed 29 August 1943 during flight testing.

V2 - Ar 234A version prototype. 2x*Jumo 004A* turbojet engines, *werk nummer 130002*, coded *DP+AW*, first flight 13 September 1943, crashed 2 October 1943 during flight testing.

V3 - Ar 234A version prototype. 2x*Jumo 004A* turbojet engines, *werk nummer 130003*, coded *DP+AX*, first flight 29 September 1943, retired August 1944.

London. The Elmers End bus garage after a direct hit by an FZG 76 on 17 July 1944.

V4 - Ar 234A version prototype. 2x*Jumo 004A* turbojet engines, *werk nummer 130004*, coded *DP+AY*, first flight 26 November 1943, retired August 1944.

V5 - Ar 234A version prototype. 2x*Jumo 004A* turbojet engines, *werk nummer 130005*, coded *GK+IV*, first flight 22 December 1943, crashed 28 August 1944 during flight testing.

V6 - Ar 234A modified to carry 4x*BMW 003A-O* turbojet engines in four individual nacelles, *werk nummer 130006*, coded *GK+IW*, first flight 8 April 1944, crashed 1 August 1944 during flight testing.

V7 - Ar 234A 2x*Jumo 004B* turbojet engines, *werk nummer 130007*, coded *GK+IX*, first flight on 22 June 1944.

London. St. Paul's Cathedral.

Hythe, England...15 August 1944.

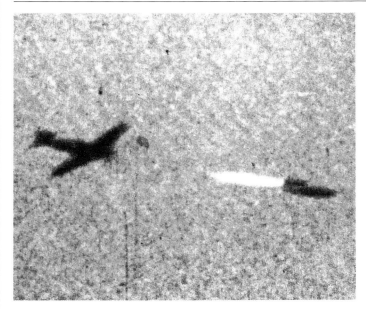

A British "Tempest" is shown closing in on an *FZG 76* over Newchurch, England.

A British pilot is shown attempting to tip the starboard wing of an FZG 76 in an effort to force it into a farm field and away from a populated area.

The British pilot was successful. This is the hole made in the farm field where the FZG 76 crashed down and its warhead exploded.

Two thousand Barrage balloons held in place by anchored steel cables covered almost 260 square miles of English countryside 31 miles long and 11 miles in width. They were successful, too, in snagging airborne FZG 76s.

Arado Ar 234C

Rockets against rockets. A British rocket battery in August 1944 near Kent attempting to bring down FZG 76s before reaching London.

V8 - Ar 234A modified to carry 4x*BMW 003A-O* turbojet engines two paired in a single nacelle. Coded *GK+IY*. Flight testing began on 1 February 1944. Dispertion at war's end is unknown.

V9 - Ar 234B werk nummer 130009, coded *PH+SQ*, and powered by 2x*Jumo 004B* turbojet engines. First flight on 5 December 1944 and available for flight testing throughout December 1944.

V10 - Ar 234B werk nummer 130010, coded *PH+SR*, and powered by 2x*Jumo 004B* turbojet engines. First flight on 7 April 1944. Destroyed on 22 July 1944.

V11 - Ar 234B werk nummer 130011, coded *PH+SS*, and powered by 2x*Jumo 004B* turbojet engines. First flight on 10 May 1944. Disposition at war's end is unknown.

V12 - Ar 234B werk nummer 130012, coded *PH+ST*, and powered by 2x*Jumo 004B* turbojet engines. First flight on 6 September 1944. Destroyed on 6 September 1944.

V13 - Ar 234B werk nummer 130023, coded *PH+SU*, and modified to carry 4x*BMW 003A-1* turbojet engines for what was hoped to be the *Ar 234C* series but known then as the *Ar 234B-1*

A battery of seven 3.7 inch cannon lined up along the English Channel at St. Leonards-on-Sea in their attempts to destroy the airborne FZG 76s and have their remains fall into the Channel rather than on London.

A night view of the rocket batteries at Romney Marsh in their attempt to destroy night flying FZG 76s.

Arado Ar 234C

Allied bombers and fighters destroyed the FZG 76's 150 foot long launch sites whereever they could find them. This one in Pas de Calais, France, was discovered in early September 1944.

series prototype. First flight on 6 September 1944 tested and reaching a record-setting 42,000 feet altitude. Destroyed 4 April 1945, and the circumstances surrounding its destruction are unknown.

V14 - *Ar 234B werk nummer 130024*, coded *PH+SV*, and powered by *2xJumo 004B* turbojet engines. First flight is thought to be in December 1944. Fate unknown at war's end.

V15 - *Ar 234B werk numer 130025*, coded *PH+SW*, and modified to carry *4xBMW 003A-1* turbojet engines. First flight on 20 July 1944. Also thought to have been flight tested with *2xBMW 003A-1* turbojet engines upon modifying their fuel regulators by replacing them with fuel regulators from the *Jumo 004B-1* turbojet engines. Reported to have been destroyed April 1945 due to unknown reasons.

V16 - *Ar 234B werk nummer 130026*, no code assigned. Research prototype to test flight performance of a wooden wing machine with a crescent wing designed by *Dipl.-Ing. Rüdiger Kosin*, *Arado's* chief aerodynamicist and powered by *2xBMW 003R* combined *003A-1* turbine and a *BMW 718* bi-fuel rocket engine providing 2,700 pounds thrust for three minutes. The *V16's* wing included a *37%* inboard sweep decreasing in two steps to *25%* outboard. The wing carried leading-edge flaps. The wing was completed in April 1945, however, the factory where the wing was constructed and stored was overrun by British ground troops. The machine is reported to have been destroyed in May 1945 due to unknown reasons.

V17 - *Ar 234 werk nummer 130027*, coded *PI+SY*, and powered by *4xBMW 003A*-1 turbojet engines. First flight on 25 September 1944. Destroyed on 4 April 1945 due to unknown reasons.

V18 - *Ar 234 werk nummer 130028*, and no coded assigned. Research prototype to test flight performance of a metal wing machine with a crescent wing and powered by *4xBMW 003A-1* turbojet engines. First flight on 8 March 1945. May have been destroyed in April 1945 to prevent it from being obtained by the Allies.

Many rail road trains loaded with new FZG 76's were destroyed by Allied fighters and bombers before arriving at their launch sites just like this one.

A rail road train load of brand new FZG 76s on their way to launch sites in France and the Low Countries.

Arado Ar 234C

Between 4 and 11 October 1944, the Luftwaffe launched *FZG 76s* from airborne *Heinkel He 111s*. It is believed that 62 *He 111s* carrying *FZG 76s* under their starboard wing lifted off at night. At least nine launched *FZG 76s* fell on London. British nightfighters believed that they shot down all the other 53 *He 111s* still carrying their *FZG 76s*.

A Heinkel He 111 has released its *FZG 76* which is now pointed toward England.

Waiting in the wings of companies such as Henschel were a wide assortment of guided missiles. Most would require airborne launching...and the *Luftwaffe* was depending upon the *Ar 234C* as their future airborne launch platform. Courtesy of *Gert W. Neumann*.

V19 - *Ar 234 werk nummer 130029*, coded *P1+WX*, and powered by 4x*BMW 003A-1* turbojet engines. First flight on 16 October 1944 and this machine is considered to be the first true *Ar 234C* prototype. Thought to still exist as of April 1945.

V20 - *Ar 234 werk nummer 130030*, coded *P1+WY*, and powered by 4x*BMW 003A-1* turbojet engines. This machine is similar to *V19* but has full cabin pressurization...the first *Ar 234C* with full cabin pressurization. First flight on 5 November 1944. Destroyed due to unknown reasons on 4 April 1945.

V21 - *Ar 234 werk nummer 130061*, coded *P1+WZ*, and powered by 4x*BMW 003A-1* turbojet engines. Redesigned cockpit cabin with a higher roof than the *Ar 234B-1* and pre-production prototype for the *Ar 234C-3* bomber version. First flight 24 November 1944. Known to still exist as of February 1945.

Arado Ar 234C

The SD 1400 or "Fri*tz* X." It had been designed by former *Messerschmitt AG* head test pilot...*Dr.-Ing. Hermann Wurster*.

A nose on view of the "*Fritz X*" on display in England post war.

A poor quality photo of the modified Fi 103 into the towed *FZG 76* trailer.

V22 - *Ar 234 werk nummer 130062*, coded *RK+EL*, and powered by 4x*BMW 003A-1* turbojet engines. Otherwise similar to *V21*. First flight on 1 January 1945. Known to still exist as of April 1945.

V23 - *Ar 234 werk nummer 130063*, coded *RK+EM*, and powered by 4x*BMW 003A-1* turbojet engines. Otherwise similar to *V21* and *V22*. First flight on 14 January 1945. Known to still exist as of March 1945.

V24 - *Ar 234 werk nummer 130064*, coded *RK+EN*, and powered by 4x*BMW 003A-1* turbojet engines. Otherwise similar to *V21*, *V22*, and *V23*. First flight on 12 January 1945. This machine is considered to be the pre-production multiple purpose type Ar 234C. Known to still exist as of March 1945.

A Luftwaffe air machine carrying a single "*Fritz X*" under its starboard wing between its piston engine and the fuselage. This launch platform was obsolete...too slow and with limited carrying abilities.

Arado Ar 234C

Arado documents suggest that their *Ar 234C* could be used to carry the manned version of the *Fi 103*. It was known as the *Fi 103R*. Shown is an *Fi 103R* shortly after release from its carrier *Ar 234C* aircraft. Digital image by *Mario Merino*.

V25 - *Ar 234 werk nummer 130065*, coded *RK+EO*, and powered by 4x*BMW 003A-1* turbojet engines. Otherwise similar to *V21*, *V22*, *V23*, and *V24*. First flight on 2 February 1945. Destroyed due to unknown reasons on 2 May 1945.

V26 - *Ar 234 werk nummer 130066* and no code assigned. Powered by 4x *BMW 003A-1* turbojet engines. Airfoil research prototype to test the features of an all-wooden wing with a thick laminar flow profile. The prototype was unfinished at war's end but nearing completion. It was purposely destroyed at war's end to prevent it from falling intact into the hands of the Soviets.

V27 - *Ar 234C-3/N*. Thought to be a night fighter research prototype and used to test a variety of air-brakes which would be used in night fighting. To have been powered by 4x*BMW 003A-1* turbojet engines. Disposition at war's end unknown.

Nose, starboard side of a Henschel *Hs 293* glider-bomb. On its wing tips were posts which supported the aerial which picked up the radio command signals from the launching aircraft, such as the *Ar 234P-5B*. Later versions, the *Hs 293D*, had a television camera control so the bombardier did not have to rely on the glider-bomb's guide flare.

Arado Ar 234C

A rear, starboard side view of a Hs 293 anti-shipping glider-bomb. The *Hs 293* entered operational service in the Summer of 1942 and performed with a fair amount of success. At the rear of the missile were flares to assist tracking by the *bombardier* in the *Ar 234P-5B*.

V28 - Ar 234C-5. Prototype for the proposed two-seat multi-purpose bomber in which the pilot and navigator/bombardier sat side by side. To have been powered by 4x*BMW 003A-1* turbojet engines. First flight scheduled for 15 April 1945. Unknown if test flown by war's end.

V29 - Ar 234C-6. Prototype for the proposed two-seat reconnaissance aircraft. To have been powered by 4x*BMW 003A-1* turbojet engines. First flight scheduled for 10 May 1945. Dispostion at war's end unknown.

V30 - Ar 234C known as *Laminarflügel III*. An airfoil research prototype with a slim swept metal wing of laminar profile. To have been powered by 4x*BMW 003A-1* turbojet engines. First flight scheduled for 25 May 1945. Unfinished as of war's end but nearing completion. It was destroyed to prevent it from falling intact to the Soviets.

V31 - Ar 234D known as Versuchsflügel IV. A prototype for a proposed reconnaissance aircraft. Initially designed to be powered by 2x*HeS 011A* turbojets, however, modified to carry 4x*BMW 003A-1* turbojet engines instead. First flight scheduled for 5 May 1945. Disposition at war's end is unknown.

V32 - Ar 234D known as *Versuchflügel V*. Similar to the V31. First flight scheduled for 15 May 1945. Disposition at war's end unknown.

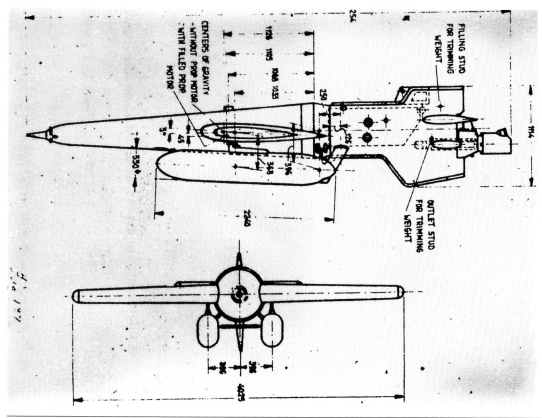

A pen and ink drawing of the Henschel *Hs 294A-0*. Two of these guided-torpedo missiles would have been carried by a standard *Ar 234C*, as well as by the *Ar 234P-5B*.

Arado Ar 234C

The Henschel Hs 294 guided torpedo missile designed by *Dr.-Ing. Herbert Wagner* of *Henschel*, as seen from above and port side. *Wagner* was a genius. In the late 1930s he had been chief of design for large aircraft...bombers at *Junkers Flugzeug*. He was not satisfied with the power of the current piston engines being built by *Junkers Motern* and was well along in designing his own operational turbojet engine. When *Dr.-Ing. Otto Mader*, head of the *Junkers Motor Works* learned that *Herbert Wagner* had been secretly researching and building prototype turbojet engines with the blessing of *Dr.-Ing. Otto Koppenberg*, Chairman of the *Junkers* Board of Directors, a huge internal fight broke out involving many powerful factions within the large *Junkers* aircraft and *Junkers* motor works. The result was that *Wagner* left *Junkers* for *Henschel*. The work *Wagner* had achieved with his turbojet engine research and prototypes were later used as the building block for the *Jumo 004* under the direction of *Dr.-Ing. Anselm Franz*.

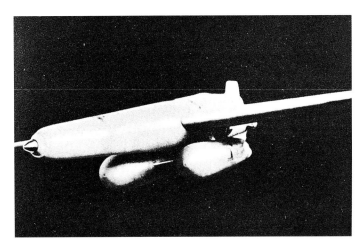

The Henschel Hs 294 air launched anti-aircraft guided missile to be carried by the *Ar 234C* and variations, as seen from its nose port side. The *Hs 294* was powered by two separate rocket motors hung beneath its fuselage.

The Henschel *Hs 298* air-to-air guided missile. This missile was powered by a solid fuel rocket engine.

Arado Ar 234C

A Focke-Wulf *Fw 200* seen on the ground with two *Henschel Hs 293* "*Egret*" air-to-ground anti-shipping guided missiles hung beneath each outboard piston engine. The "*Condor*" was too slow and poorly defended to be an effective missile launch machine.

V33 - *Ar 234D* and prototype for the *Ar 234D-1* bomber. and Powered by 2x*HeS 011A* turbojet engines. Scheduled completion about 15 August 1945. Disposition at war's end unknown.

V34 - *Ar 234D*, a prototype for the proposed *Ar 234D-2* night fighter and to have been powered by 2x*HeS 011A* turbojet engines. Disposition at war's end is unknown.

V35 - *Ar 234D*, a prototype for the proposed *Ar 234D-3*. Projected date of completion was 20 September 1945. Disposition at war's end is unknown.

V36 - *Ar 234D-1a* bomber prototype. To have been powered by 2x*HeS 011A* turbojet engines. Scheduled date of completion was 10 October 1945. Disposition at war's end is unknown.

Left to right: *Walter Blume (Arado), Bruno Loerzer (RLM), Hermann Göring (RLM), and General Erhard Milch (RLM)*.

A Heinkel *He 177A-3* illustrated with two *Henschel Hs 293* "*Egret*" air-to-ground anti-shipping guided missiles suspended from missile racks. The *He 177* had the speed, range, and carrying ability but it was horribly unreliable due to numerous mechanic problems stemming from its twin coupled *DB 610* engines.

Arado Ar 234C

Rüdiger Kosin, **former chief aerodynamitist at** *Arado*. *Kosin* **was given the duties of modifying the twin turbojet engined** *Ar 234B* **into the four turbojet engined** *Ar 234C*. **Photographed by the author. Munich, 1986.**

V37 - *Ar 234D-2* night fighter prototype. To have been powered by 2x*HeS 011A* turbojet engines. Scheduled date of completion was 1 November 1945. Disposition at war's end is unknown.

V38 - *Ar 234D-3* prototype with unspecified duties. To have been powered by 2x*HeS 011A* turbojet engines. Scheduled date of completion was 20 November 1945. Disposition at war's end is unknown.

V39 - *Ar 234D-1b* bomber prototype. To have been powered by 2x*HeS 011A* turbojet engines. Scheduled date of completion was 10 December 1945. Disposition at war's end is unknown.

V40 - *Ar 234D*-1*c* prototype. To have been powered by 2x*HeS 001A* turbojet engines. Scheduled date of completion was 31 December 1945. Disposition at war's end is unknown.

Several different uses were planned for the production *Ar 234C* and included the following:

• **Ar 234C-1** - reconnaissance duties - Single-seat cockpit and the pilot flew the machine in a pressurized cabin, powered by 4x*BMW 003A*-1 turbojet engines producing 1,760 pounds of thrust each at sea level. In addition, *Rb 75/30, Rb 50/30* or *Rb 20/30* photo reconnaissance cameras would be installed similar to those used in the *Ar 234B-1*. Armament included 2x*MG 151* 20 mm rearward-firing cannon with 250 rounds each and mounted in a ventral position (blister) beneath the fuselage, about midway between the cockpit and the tail assembly.

Kosin's **job,** *al***ong with his assistants, was to make the basic** *Ar 234B* **into a real "***Engländer Bomber***," that is, carrying an effective bomb load to London on a regular basis without fear of being intercepted. The** *Ar 234B* **could run photo reconnaissance over England without being intercepted.** *Hilter* **wanted a dedicated bomber version to do the same. Water color by Loretta Dovell.**

Arado Ar 234C

Specifications

Wing span:	46 feet 3 1/2 inches
Wing area:	284.167 square feet
Length, overall:	41 feet 5 1/2 inches
Height:	14 feet 1 1/4 inches
Weight, empty:	5,400 pounds
Weight, loaded:	13,200 pounds
Weight, maximum loaded:	20,600 pounds
Fuel load:	815 gallons
Speed, maximum at sea level:	515 mph
Speed, maximum at 19,700 feet:	542 mph
Speed, landing:	99.3 mph
Time to climb to 19,700 feet altitude:	11.9 minutes
Range:	920 miles
Crew:	1

A rare sight. Shown is a lone twin piston engined *He 111* over central London on 9 September 1940. The four turbojet engined *Ar 234C* was expected to do this on a regular basis.

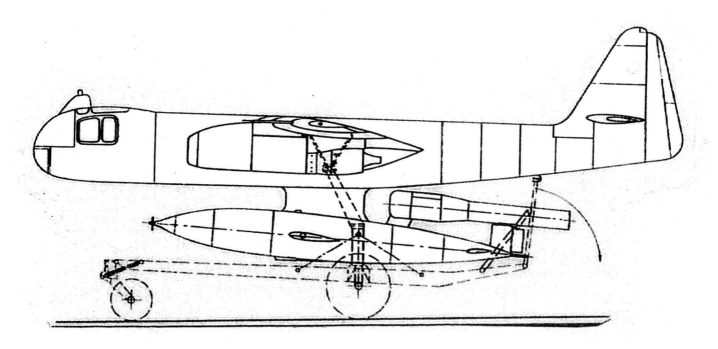

Hilter wanted *E*ngland to be attacked again and again by his *Fi 103* flying bomb. Initially, this would be the first duties of the *Ar 234C*...to serve as a launch platform for the *Fi 103s*. The pen and ink drawing of the *Ar 234C* is shown carrying an *Fi 103* flying bomb beneath its fuselage, and the whole arrangement is fitted to a three-wheeled take-off trolley. The take-off trolley would detach upon lift off.

Arado Ar 234C

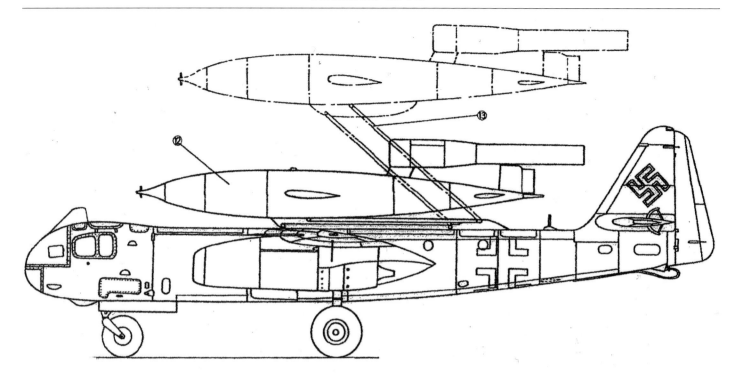

An Arado company pen and ink drawing featuring an *Ar 234C* carrying an *Fi 103* flying bomb (*FZG 76*) up on its back. The carrying apparatus was composed of a series of extendable hydraulic arms to lift the bomb and fuel heavy *Fi 103* above the *234C's* tail assembly so it could be safely released.

- **Ar 234C-2** - offensive bombing duties - The machine with a configuration similar to the *Ar 234C-1*, but without cannon and carrying 1x2,205 pound bomb or 2x1,100 bombs. Single-seat cockpit. It is reported that this machine was able to lift-off within 3,000 feet (actually 2,916 feet), however, the takeoff run could be further reduced to 2,001 feet through the use of 2xHWK 501 takeoff assist 1,000 pound thrust rockets known as *Rauchgeräte* units.

Specifications

Wing span:	46 feet 3 1/2 inches
Wing area:	284.167 square feet
Speed, maximum at mean weight of 17,000 pounds:	555 mph
Speed, landing:	97 mph with a mean landing weight of 13,100 pounds
Weight, loaded:	22,000 pounds
Range:	NA
Time to climb to 33,300 feet attitude carrying a 4,410 pound [2,000 kilogram] bomb load:	17.8 minutes
Takeoff run:	2,916 feet
Takeoff run with 2x1,102 pound [2x500 kilogram] thrust from HWK 500 rocket assist units:	2,001 feet.
Crew:	1

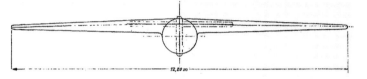

The *E.377's* overall wing span was to have been 35 3/4 feet [10.90 meters].

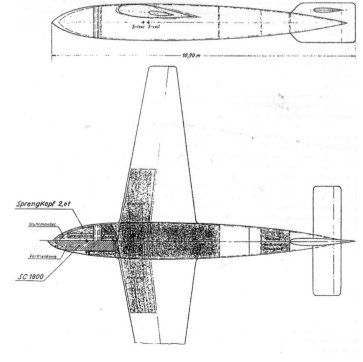

Arado Ar 234C

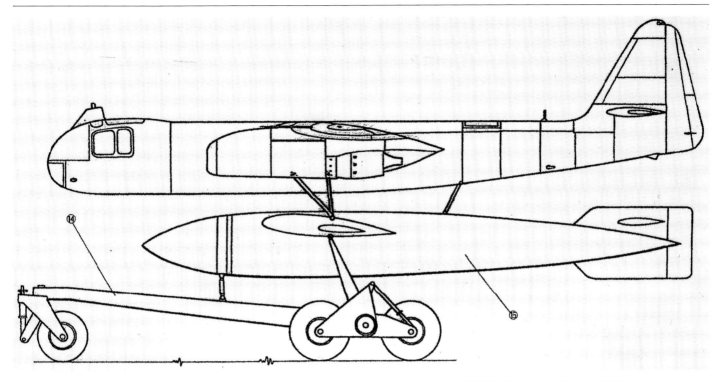

Arado company documents suggest that their *Ar 234C* could be used to carry one of their *Ar E.377* glide bombs. The *Ar E.377* was to have been similar to the *Fi 103*, however, without the *Schmidt-Argus* pulse-jet engine. Its tail assembly would have been modified, too. The circled numbers refer to: #14 - Three-wheeled takeoff dolly, and #15 - *Ar E.377* glide bomb.

• **Ar 234C**-3 - multi-purpose duties - such as a bomber, night fighter, or ground attack aircraft. All versions were to have single-seat cockpits and powered by 4x*BMW 003A-1* turbojet engines producing 1,760 pounds thrust each. One *Arado* report stated that these machines could also be powered by 2x*Jumo 004B* turbojet engines. In addition, *Arado* engineers were planning to install the *BMW 003D* with its promised 2,420 pounds thrust at sea level as opposed to the standard *BMW 003A-1* and *C*-version with its 1,760 pounds and 1,980 pounds thrust, respectively. The *BMW 003D* turbojet engine was never built.

Specifications

Wing span:	46 feet 3 1/2 inches [14.44 meters]
Wing area:	284.167 square feet
Length, overall:	[12.66 meters]
Height, overall:	[4.20 meters]
Weight, empty:	14,400 pounds [6,500 kilograms]
Weight, loaded:	24,250 pounds [11,000 kilograms]
Speed, maximum:	496 mph [800 km/h] at sea level, or 530 mph [852 km/h] at 19,685 feet [6,000 meters] altitude
Speed, maximum at 18,000 feet altitude on return flight:	555 mph
Time to climb to 32,810 feet [10,000 meters] altitude:	16.7 minutes
Service ceiling:	36,092 feet [11,000 meters]
Range, maximum:	765 miles [1,215 kilometers]
Armament:	2x*MG 151* 20 mm cannon
Power:	4x*BMW 003 A-1* turbojet engines providing 1,760 pounds [800 kilograms] of thrust each
Crew:	1

Left: An *Ar E.377* glide bomb. This bomb also called for the *Ar 234C* as its launching platform. Its fuselage overall length was 35 3/4 feet [10.90 meters]. In its nose was the same explosive charge as found in the *SC 1800* bomb.

Arado Ar 234C

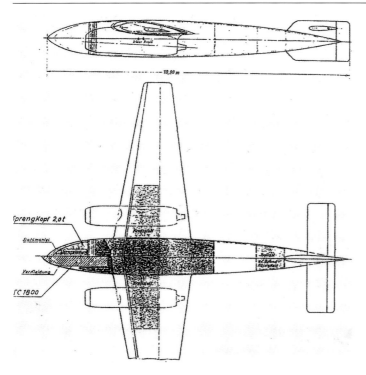

Another version of the *Ar E.377* was to have been powered by 2x*BMW 003A-1* turbojet engines. Although turbojet powered, this second *E.377* version would have had the same overall dimensions of the glide version.

Former Arado chief aerodynamitist *Rüdiger Kosin* (left) and *Wilhem Benz* (right) designer formerly of *Heinkel AG. Kosin*, in addition to developing the *234B* into the *234C*, was working on wings with various degrees of sweep-back and shapes at war's end for planned variations of the *Ar 234C. Benz* had worked with *Walter Günter* and others at *Heinkel* in designing the proposed *HWK 509C* powered *He P.1077* "*Julia*." Earlier, in the late 1930s, he had been involved in *Heinkel's* bi-fuel rocket powered *He 176* aircraft, too. Photographed by the author in Munich in the mid 1980s.

• **Ar 234**C-4 - reconnaissance duties - Powered by 4x*BMW 003A-1* turbojet engines producing 1,760 pounds thrust each at sea level. Single seat cockpit. Armament was to include 2x*MG 151* 20 mm cannon with 250 rounds each.

Specifications

Speed, maximum:	547 mph at 1,6500 feet altitude and 525 mph at 33,300 feet altitude
Weight, mean flying:	20,000 pounds
Endurance:	2 hrs 10 mins with 1,680 gallons of fuel 1 hr 55 mins with 945 gallons of fuel 1 hr 45 mins with 815 gallons of fuel

• **Ar 234**C-5 - high speed bombing duties - Powered by 4x*BMW 003A-1* turbojet engines producing 1,760 pounds thrust each at sea level. This was the first machine to have a two-seat cockpit arrangement with the pilot and navigator/bombardier sitting in staggered seats. A prototype was constructed and it is known as the *V28*.

Specifications
None available.

Dipl.-Ing. Wilhem Benz's proposed bi-fuel liquid rocket-powered fighter/interceptor in powered flight: the *Heinkel He P.1077* "*Julia*." Scale model and photographed by *Jamie Davies*.

• **Ar 234**C-6 - long-range reconnaissance duties - Powered by 4x*BMW 003A-1* turbojet engines producing 1,760 pounds thrust each at sea level. A prototype was constructed and it is known as the *V29*.

Specifications
None available.

Arado Ar 234C

Arado had its own version of a miniature fighter which would be carried aloft via a carrier aircraft...an *ArE.381*. The pilot of the *Ar E.381* was to have flown this tiny interceptor in the prone position. For the pilot's protection, the lower portion of its nose was heavily armored with 5 mm steel plate. The *Ar E.381* is seen here from its nose port side. Scale model and photographed by *Günter Sengfelder*.

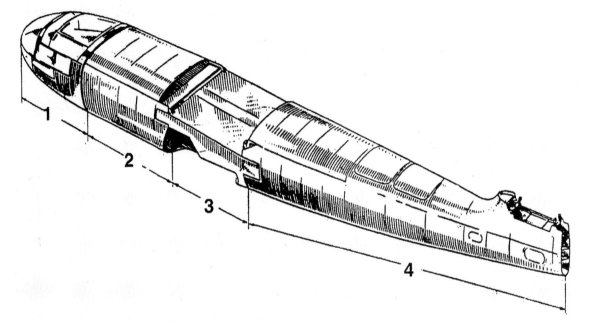

An Arado company pen and ink drawing of the *Ar 234C's* fuselage. Notice that the fuselage has been divided into four sub-sections: 1. Nose/cockpit; 2. Forward fuselage (between cockpit and wing); 3. Center section (wing attachment section including main gear storage); and 4. Rear or aft fuselage.

Rüdiger Kosin told this author that *Walter Blume* instructed him to take the basic *Ar 234B's* fuselage, wings, and tail assembly and modify it to accept four *BMW 3302* turbojet engines with their 1,102 pounds [500 kilograms] of thrust each. As to where to hang the four turbojet engines under the wing, *Kosin* and his group had to decide for themselves. They weren't sure...two engines in a single nacelle under each wing, or hang them individually. So *Kosin* and his group decided to build two prototypes: one with paired engines and the other with individual engines.

The port side view of Arado's proposed miniature fighter known as the *Ar E.381*. Scale model and photographed by *Günter Sengfelder*.

Arado Ar 234C

The restored fuselage of NASM's *Ar 234B-2* shown at their Garber Restoration Facility, Silver Hill, Maryland, helps us better visualize the four sections of the *Ar 234B's* fuselage *Kosin* and his group had to work with as seen in the previous photograph.

• **Ar 234C-7** - night fighting duties - To be powered by 4x*BMW 003A-1*, or 2x*HeS 011A*, or 2x*Jumo 004D* turbojet engines. It appears from *Arado* documents that the 2x*HeS 011A's* with their 2,860 thrust each would have been the engines of choice for this machine. Armament was to be 1x*MG 151* 20 mm cannon with 500 rounds and 2x*MK 108* 30 mm cannon with 100 rounds each located in a ventral enclosure beneath the fuselage. All rounds for the 2x*MK 108* 30 mm cannon were to be contained in this ventral pack under the center part of the fuselage. Total fuel included 815 gallon tank within the fuselage and two drop tanks holding 132 gallons each hung beneath each inboard turbojet engine nacelle. This machine would have been equipped with the highly successful *Telefunken FuG 350 "Naxos"* homing-in" system.

Specifications
None available.

• **Ar 234C-8** - bombing duties - Single seat machine with the pilot inside a pressurized cabin. This sub-type was to have been powered by 2x*Jumo 004D* turbojet engines producing 2,310 pounds thrust each at sea level. Armament included 2x*MG 151* 20 mm forward-firing cannon with 250 rounds each. Normal bomb load was 2,205 pounds[1,000 kilograms].

Specifications
None available.

The upper wing surface of NASM's *Ar 234B-2* after restoration. The wing on the *Ar 234C* was very similar. The wing's leading edge is seen resting on the dolly.

Arado Ar 234C

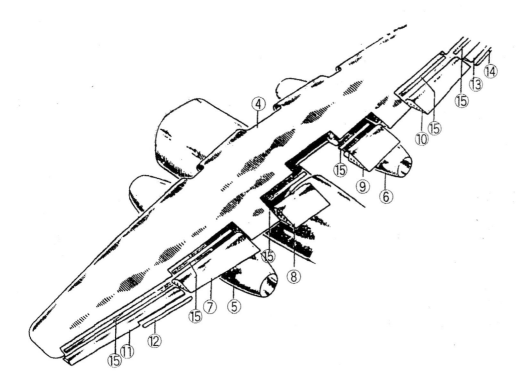

The wing components of the Ar 234B before *Kosin* and his team converted it into the *Arado Ar 234C*:
#04 - wing assembly
#05 - single port turbojet engine
#06 - single starboard turbojet engine
#07 - port outboard landing flap
#08 - port inboard landing flap
#09 - starboard inboard landing flap
#10 - starboard outboard landing flap
#11 - port aileron
#12 - port aileron trim tab
#13 - starboard aileron
#14 - starboard aileron trim tab
#15 - port aileron leading edge cover

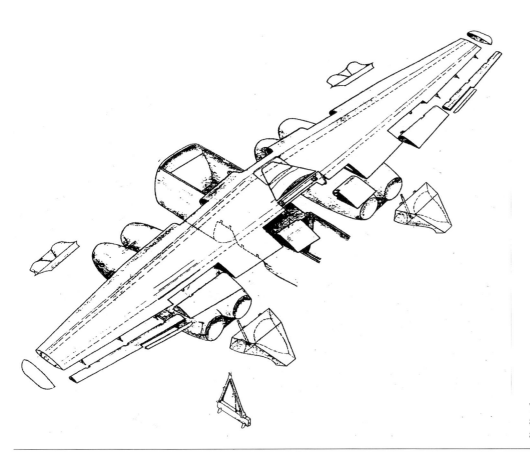

The former *Ar 234B* wing assembly after conversion by *Kosin* into a wing assembly for the *Ar 234C*.

Arado Ar 234C

- **Ar 234CV1** - duties unknown - Powered by 2x*Daimler-Benz DB 007* turbojet engines producing 2,811 pounds of thrust at sea level. This engine, known today as a fan jet or jet by-pass engine, had been abandoned by the *RLM* in 1944 due technical difficulties and *Daimler-Benz* engineers assigned to help out *Heinkel-Hirth*, especially the *HeS 011A*.

Specifications
None available.

- **Ar 234CV2** - *Deichselschlepp* or fuel trailer - This machine was intended for long-range duties and in order to do so the *Ar 234C* would tow an expendable long-range tank carrying 880 gallons. This tank would be attached to the *Ar 234C* by means of a semi-rigid tube which served as a tow bar and fuel feed pipe.

Specifications
None available.

- **Ar 234CV3** - towing a *Henschel Hs 294* anti-ship torpedo missile - This sub-type would have towed a *Hs 294* air-launched torpedo guided missile. The aircraft was to have been powered by 4x*BMW 003A-1* turbojet engines producing 1,760 pounds thrust each at sea level.

An *Ar 234C-3* featuring its full wing span assembly. The wing had been modified from the twin engined *Ar 234B*. Scale model and photographed by *Günter Sengfelder*.

The rear port side of a *BMW 003A-0* turbojet engine of the type and appearance used to power most of the *Ar 234C* series machines. To the right in the photo is the *003A-0's* tapered adjustable tail cone. Notice the absence of vertical fins on the tail cone. The initial *BMW 003* turbojet engine was known as the *BMW/Bramo 3302*. It had a 6-stage compressor, however, its thrust output failed to meet specifications and was abandoned. *BMW* had to completely redesign the engine around a 7-stage compressor and even afterward, they continued to have problems failing to meet the expected thrust of 1,764 pounds.

Arado Ar 234C

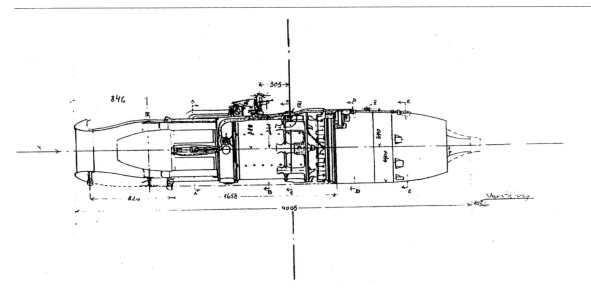

Pen and ink drawing featuring the port side of a *BMW 003A-0* turbojet engine.

Specifications

Speed, maximum:	470 mph at sea level and 510 mph at 26,250 feet altitude
Weight, loaded:	22,000 pounds
Weight of 1x*Hs 294* torpedo/missile:	4,190 pounds
Rate of climb:	3,050 feet per minute at sea level 1,800 feet per minute at 26,250 feet altitude
Range:	250 miles at sea level or 475 miles at 26,250 feet altitude

A ground level view of the *Ar 234 V8* mounted on its three-wheeled takeoff dolly. Notice that its outriggers extend down from under the outer turbojet engine's nacelle.

A ground level nose on view of the *Ar 234 V6*. Notice the short outrigger arms coming out of the lower cowling on the two outboard *BMW 003A-0* turbojet engines. This machine's maiden flight occurred on 25 April 1944, and it required the use of a three-wheeled dolly as shown.

Arado Ar 234C

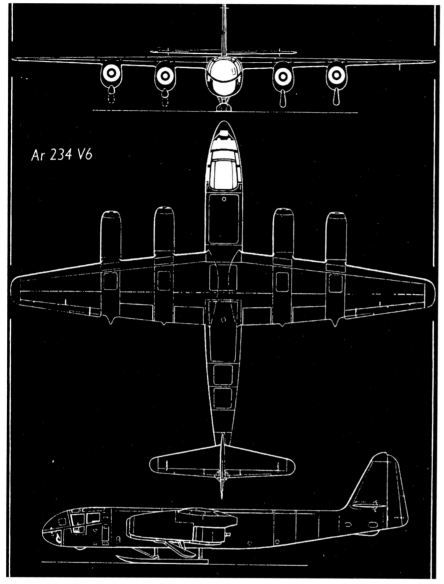

A pen and ink 3-view drawing featuring the Ar 234 V6, coded *GK+IW*, werk nummer 130006, showing its individually mounted 4x*BMW 003A-0* turbojet engines.

- **Ar 234CV4**- towing a *Fieseler Fi 103* flying bomb - The *Fi 103* was to have been mounted on a trailer comprising a fixed undercarriage with a small center-section and wings, the whole unit being attached to the parent towing aircraft by a towing arm. It was proposed that the undercarriage be jettisoned after take-off and the tow bar detached when the *Fi 103's* pulse jet engine had been started and the bomb directed at its target. It is believed that this machine would have also towed to altitude the manned *Fi 103R* or piloted flying bomb.

Specifications
None available.

- **Ar 234CV5** - carrying a *Fieseler Fi 103* flying bomb or the piloted *Fi 103R* - each *Fi 103* version would have been carried on a cradle on the back of the *Ar 234CV5*. Launching the flying bomb/piloted bomb, would be accomplished by raising the *Fi 103* in its cradle by means of hydraulically operated arms in order to clear the top of the parent aircraft.

- **Ar 234D-1** - reconnaissance duties - This machine was to have been powered by 2x*HeS 011A* turbojet engines producing 2,860 pounds thrust each at sea level. The *HeS 011A* was never released for field testing prior to the surrender.

Specifications
None available.

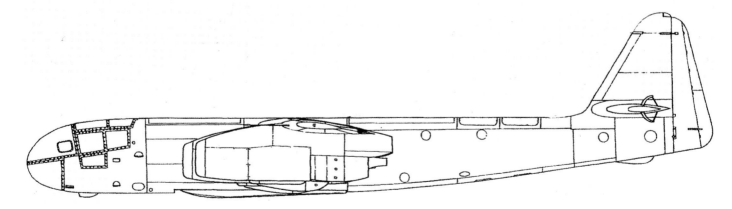

A pen and ink drawing from Arado featuring the port side of their *Ar 234V6*.

Arado Ar 234C

The *Ar 234 V6* as seen from its nose starboard side. Four men in black jumpsuits (ground crew) appear to be manhandling...moving the machine by the tow bar attached to its three-wheeled dolly.

• **Ar 234D-2** - bomber duties - To be powered by 2x*HeS 011A* turbojet engines producing 2,860 pounds thrust each at sea level. Other wise similar to the *Ar 234D-1*.

Specifications
None available.

• **Ar 234D-3** - specific duties unknown - To be powered by 2x*HeS 011A* turbojet engines producing 2,860 pounds thrust each at sea level. *Arado* documents are unclear regarding this type, however, it is thought to be similar to the *Ar 234D-1* and *Ar 234D-2*.

Specifications
None available.

• **Ar 234P-1** - night fighting duties - This machine would have carried a two man crew. It would have had a lengthen fuselage nose to accommodate a *Telefunken FuG 240* "centimetric" or 9 centimeter intercept radar and thereby increasing the fuselage overall length to 43 feet 6 inches from the typical *Ar 234C* with an overall length of 41 feet 5 1/2 inches. Powered by 4x*BMW 003A-1* turbojet engines producing 1,760 pounds thrust each. Armament included 1x*MG 151* 20 mm cannon with 300 rounds and 1x*MK 108* 30 mm cannon with 100 rounds.

Specifications
None available.

The *Ar 234 V6* prototype with four individually mounted turbojet engines. There are rumors that this machine actually was test flown on several reconnaissance missions. Very unusual for a prototype machine but these were becoming desperate times for Germany. Notice that this *V6* prototype uses a three-wheel dolly for takeoff and a built-in skid arrangement for landing.

Arado Ar 234C

All of the *Ar 234 V6's* four *BMW 003A-0* turbojet engines apparently have been started, and the machine waits at the end of the runway.

Arado's turbojet engine specialists appear to be starting the outer port side *BMW 003A-0* turbojet engine. They are using some sort of portable auxiliary starter engine with a drive-shaft extended right into the engine where its *Riedel* starter normally would be.

• **Ar 234P-2** - night fighting duties - Similar to the *Ar 234P-1*, but more heavily armored with cockpit metal plating fabricated out of 13 mm thick steel sheet. This machine would have been fitted with the *Telefunken FuG 360 "Naxos"* homing in radar system.

Specifications
None available.

• **Ar 234P-3** - night fighting duties - Two man crew. Powered by 2x*HeS 011A* turbojet engines producing 2,860 pounds thrust each. Armament included 2x*MG 151* 20 mm cannon with 125 rounds each and 2x*MK 108* 30 mm cannon with 100 rounds. All cannon were fixed forward.

Specifications
None available.

Arado turbojet engine specialists appear to be starting its inside starboard *BMW 003A-0* turbojet engine through the use of their portable auxiliary starter.

Arado Ar 234C

The *Ar 234 V6* appearing at the edge of the runway, perhaps making pre-flight preparations prior to takeoff. From this view the four engine-mounted outriggers can be clearly seen, as well as the machine's fuselage-mounted landing skid.

• **Ar 234P-4** - night fighting duties - Same as the *Ar 234P-3* except this machine was to have been powered by 2x*Jumo 004D* turbojet engines producing 3,060 pounds thrust each at sea level.

Specifications
None available.

• **Ar 234P-5A** - night fighting duties - Three man crew. Powered by 2x*HeS 011A* turbojet engines producing 2,860 pounds thrust each. Reduced fuel carrying capacity from a mean of 815 gallons to 660 gallons. Armament included 1x*MG 151* 20 mm cannon with 300 cannon rounds and 2x*MK 108* 30 mm cannon with 100 rounds each. These cannon fired forward. In addition, 2x*MK 108* 30 mm with 100 round each were fixed to fire upward at an oblique angle. This machine would have been fitted with the *Telefunken FuG 240* "*Berlin N-la*" 9 centimeter wavelength "centimetric" interception radar. It would also had the *FuG 350* "*Naxos*" homing-in" system.

The parachute-holding box for the *Ar 234C's* three-wheel takeoff dolly.

Specifications
None available.

A nose port side view of the three-wheeled takeoff dolly supporting the *Ar 234 V6*.

The *Ar 234 V6's* three-wheeled dolly takeoff as seen from its rear port side.

Arado Ar 234C

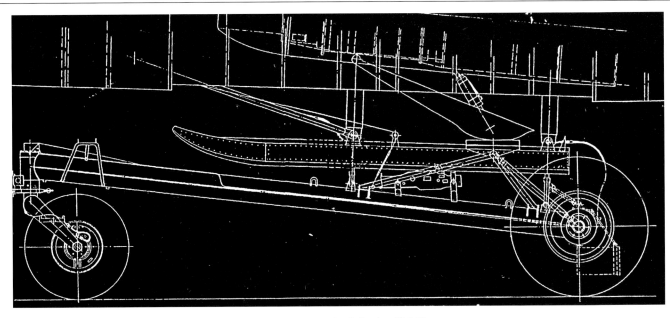

A pen and ink drawing of the port side of the *Ar 234 V6's* three-wheeled takeoff dolly.

• **Ar 234P-5B** - *AWACS* duties - Two or three man crew. Powered by 4x*BMW 003A-1* turbojet engines producing 1,760 pounds thrust each. The outstanding feature of this machine was its dorsal-mounted *AWACS* or airborne warning and control system. A prototype of *AWACS* revolving disk was under development and its shape had been wind tunnel tested. As an *AWACS* machine, this *Ar 234P-5B* would have provided all weather surveillance, command, control, and communications as needed by commanders of *Luftwaffe* defense forces.

Specifications
None available.

• **Ar 234R** - reconnaissance duties - Powered by a single *HWK 109-509* bi-fuel liquid rocket engine of 4,410 pounds [2,000 kilogram] thrust. It is believed that this *Ar 234R* (for rocket powered) would be towed up to about 26,248 feet [8,000 meters] altitude by a *Heinkel He 177A* and then the *Ar 234R's HWK 509* would be engaged allowing the machine to propel itself up to its operating altitude. On the other hand, this *Ar 234R* might have scheduled to use two or more *BMW 003R* engines. These engines consisted of a *BMW 003A-1* turbojet producing 1,760 pounds thrust plus an attached BMW *718* bi-fuel liquid rocket engine which produced 2,750 pounds thrust for 3 minutes and a combined thrust of 4,510 pounds at sea level.

Specifications
None available.

A full starboard side view of the *Ar 234 V6*. Notice the extended outriggers on the machine's outer starboard *BMW 003A-0* turbojet engine.

Arado Ar 234C

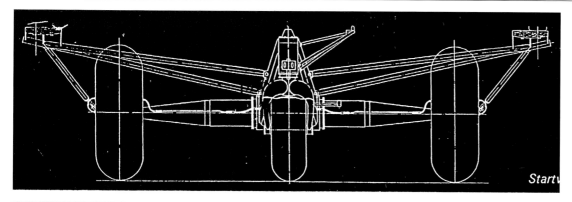

A pen and ink drawing featuring the front of the *Ar 234 V6's* three-wheeled takeoff tolly. The machine's two inside *BMW 003A* turbojet engines rested on the dolly's outer supports.

Number of *Ar 234Cs* Built Prior to the Surrender

It is thought that approximately 10 *Ar 234C* flight-ready prototypes were constructed prior to Germany's surrender on 8 May 1945. In addition, 14 *Ar 234C-3* pre production multi-purpose versions were completed. Also, several *Ar 234C-1* reconnaissance reversions were completed, however, they may have been included in the 14 multi-purpose *Ar 234C-1s* constructed. An unknown number of *Ar 234C* types and derivatives were under construction. So, for all practical purposes, about a total of 25 *Ar 234Cs* of all types had been constructed and used in various flight testing purposes.

Number of *Ar 234s* Surviving in the Year 2000

No *Ar 234Cs* are known to have survived beyond the surrender, although this author has a photograph of several U.S. Army soldiers standing near the starboard nose of an *Ar 234C*. This machine appears in the photograph to be in fairly good condition. Its post-war fate is unknown. In fact, only one known *Ar 234B* survives today, and it is a *234B-2*. The sole example of *Walter Blume's* pioneering machine left in the world is at the National Aeronautics and Space Museum (NASM), Washington, D.C. It is an *Ar 234B-2, werk numer 140312,* and it has been faithfully and beautifully restored and is on indoor public display. It is painted in the markings of *KG 76* and coded *F1+GS*. These marking and code were not found on the machine at the time of its capture in Norway, however, NASM officials believed it represented typical markings for a machine of

A close up view of the *Ar 234 V6's* port side under the outboard turbojet engine nacelee retractable outrigger.

A poor quality photograph of the port rear side of the *Ar 234V6* sitting atop of its three-wheeled takeoff tolly appearing ready for take off.

Arado Ar 234C

This is a photograph of an *Ar 234B* moments after becoming airborne. It was the same, too, for the *Ar 234 V6*. The nose of its three-wheeled dolly is in the up position, too, as momentum keeps it moving along a few feet below the turbojet engined *Arado*.

The *Ar 234B* continues to slowly climb while its three-wheeled dolly's breaking parachute is unfolding. Eventually, the dolly will be brought to full stop by its parachute.

It is time for landing the *Ar 234B*. The procedure and method was the same for the four-engined *Ar 234 V6*. Notice that the fuselage-mounted skid is fully extended underneath the fuselage. Its outrigger arms are not clearly visible.

Arado Ar 234C

A starboard side view of the *Ar 234B* as it has flared out during its final landing approach only inches above the grass.

Touch down on the grass. Notice the dust cloud, caused by the landing skid, trailing behind the machine.

This *Ar 234B* appears as if it is resting on the grass...but its not. When the machine was down and supported by its fuselage landing skid and outriggers, it was only inches above the grass.

Arado Ar 234C

Ar 234V3 is about to touch down...not exactly level, with its port wing slightly down.

this type. A war's end, this machine surrendered to the RAF at Stavanger, Norway, and was handed over to the USAAF. Later, USAAF *Colonel Harold Watson,* leader of a group of American pilots ferrying former *Luftwaffe* machines to shipping ports in France and nicknamed the *"Watson's Whizzers,"* flew the machine to Melun and then on to Cherbourg, France, where it was loaded on *HMS Reaper* for shipment to the USA. Upon arrival in the United States, the machine was transported to Freeman Field, Indiana, and made flight ready about May 1946. It is reported to have been test flown for over twenty hours before it was retired and placed in storage at Park Ridge, Illinois, in 1947. Several other *Ar 234Bs* brought to the United States were not so fortunate when their testing days were over. Most were used as landfill at the end of the runway at the Patuxent River Naval Air Station, Patuxent River, Maryland, in the 1960s when the Air Station's runway was lengthened.

So ends *Arado's* attempt to convert their successful twin turbojet engined *Ar 234B-1* into a four turbojet engined machine known as the *Ar 234C*, which among its other intended duties, was hoped

Easy does it! The *Ar 234V3's* nose has rubbed along the snow-covered landing strip. Difficult job landing on skids...under all circumstances.

to be a highly successful *"Die Engländer Bomber"* to meet the demands of their long-suffering *Führer* and his mania for revenge against England. This was a machine that could have done it, too, and with great effectiveness had the war continued on for another year.

The *Ar 234V3* shown here, but this time fitted for the first time with a braking parachute. Again, this landing is not exactly level with the starboard wing down near the grass.

Arado Ar 234C

The *Ar 234V6* seen after crash landing in a plowed field. Piloted by *Arado* test pilot *Ubbo Janssen*, 1 June 1944.

A poor quality photograph of the *Ar 234V6* from behind after its crash landing in a field on 1 June 1944. Notice that all four of its landing flaps are in the full down position. Barely visible left of the white blotch is a darken furrow created by the machine's center landing skid as it dug in the soft dirt.

The *Ar 234V6* sitting out in the field on 1 June 1944, and seen from its starboard rear side.

Arado Ar 234C

A fully loaded *Ar 234* required takeoff assistance. This was usually provided by 2x*Walter Ri202 (HWK 500A1) RATO* units. Each *Walter Ri 202* provided 1,102 pounds [500 kilograms] of thrust for 30 seconds. Immediately after burnout, the *RATO* units were released and floated down to the ground via its own parachute.

An *Ar 234* takes to the air with the help of 2x*Walter Ri202 RATO* units. Notice this *234* used a three-wheeled takeoff dolly. The dolly has been released, and it is being brought to a stop by its attached parachute. The *Ar 234C* would have required the use of *RATO* units, also.

This *Ar 234B-2* "blitz bomber" is taking off with the aid of two *HWK 500* rocket assist takeoff bi-fuel rocket engines. Notice the single *SC 1000* bomb beneath its fuselage.

Arado Ar 234C

The *Ar 234* continues to climb with the help of 2,204 pounds thrust from its 2x *Walter Ri202 RATO* units.

After 30 seconds the *Walter Ri202 RATO* units had completed their burn. They were released by the pilot and fell back to earth with the help of a parachute as seen in this photograph of an *Ar 234* successfully airborne.

Continuing to climb the *Ar 234B's HWK 500* takeoff assist rockets are seen putting out dense, black smoke...their signature.

Arado Ar 234C

The *Walter RATO* is shown mounted next to the *BMW 003B* turbojet engine on *Ar 234 V9*. The bag-like item on its nose is its packed parachute.

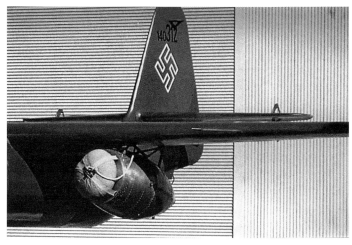

Another view of the *Walter RATO* unit shown here on the port wing of *NASM's Ar 234B-2*.

The *Ar 234 V8* on its three-wheeled takeoff dolly...just as its sister ship...the *Ar 234 V6* had been equipped for its flight testing.

Right: A pen and ink three view drawing of the *Ar 234 V8*. The machine was pretty much the same as its *V6* sister, however, *Arado* engineers placed two *BMW 003* engines in a single nacelle/pod under each wing...port and starboard.

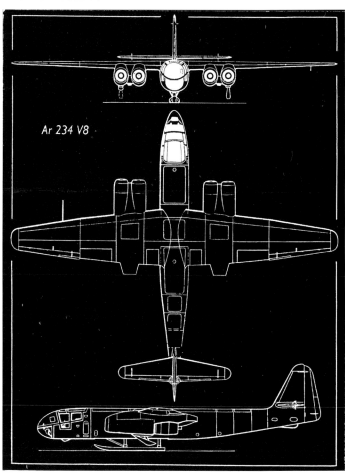

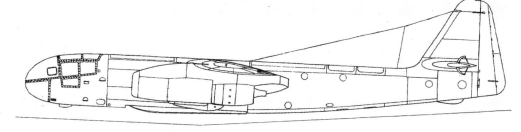

A pen and ink drawing from *Arado* featuring the port side of their *Ar 234V8*.

Arado Ar 234C

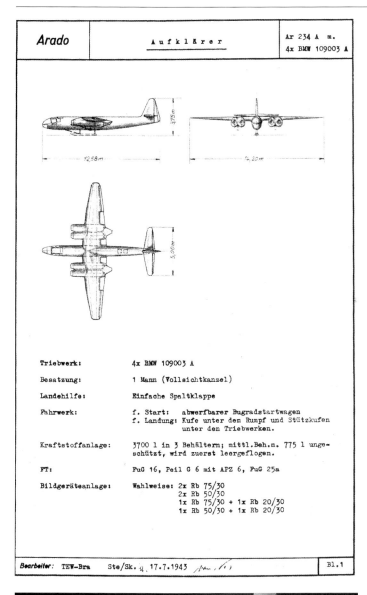

A wooden mockup of an *Ar 234C* and featuring its paired starboard side *BMW 003A-0* turbojet engines and the cowling which covered them.

Left: The first page of a two page *Arado* company document dated 17 July 1943 and featuring its *Ar 234 V8* on its landing skid and outriggers. The document suggests that the machine is to be a photo reconnaissance aircraft.

Left: A port side view of the *Ar 234 V8* and featuring its port side 2x*BMW 003A-0* engine nacelle. The machine looks fairly complete and only lacking a paint job. Its maiden flight occurred on 4 February 1944. Although it is reported to logged about 2 1/2 hours of flight time before being retired, it was enough to suggest that placing two *BMW 003A-0's* together in a single nacelle was better aerodynamically than the four separate turbojets tried on the *Ar 234 V6*. Right: Some of the *BMW 003A-0's* had two vertical fins on their adjustable tail cone. The ones which did, looked like this pair of *003A-0s* in the photo. The *Ar 234C* shown here appears to be a mockup but the engines are real.

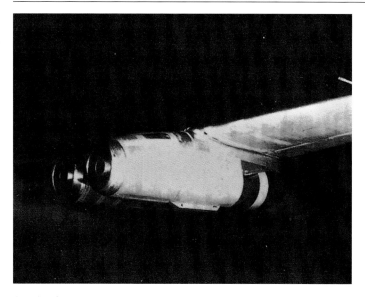

A pair of *BMW 003A-0* turbojet engines on the *Ar 234C* mockup as seen from its nose starboard side at ground level.

An inboard view of *2xBMW 003A-0* turbojet engines installed under the starboard wing of an *Ar 234C*. It appears that these engines have been rotated a few degrees to port during the installation in order to gain a lower profile. However, the installation looks neat, orderly, and functional, although it must have been a bit difficult at times to gain access to the various components.

The standard *BMW 003A-0 7* compressor stage turbojet engine which was capable of producing 1,760 pounds thrust at sea-level. Each complete engine weighed 1,342 pounds. It had a diameter of 2 feet 3 inches and was 10 feet 4 inches long.

Arado Ar 234C

2x*BMW 003A-0's* as seen on the starboard wing of *Ar 234C V19*.

2x*BMW 003A-0's* as seen on the port wing of *Ar 234C V19*. Notice its bloated appearance aft its air intake ring. This was necessary to accommodate the oil tanks mounted on the air intake duct.

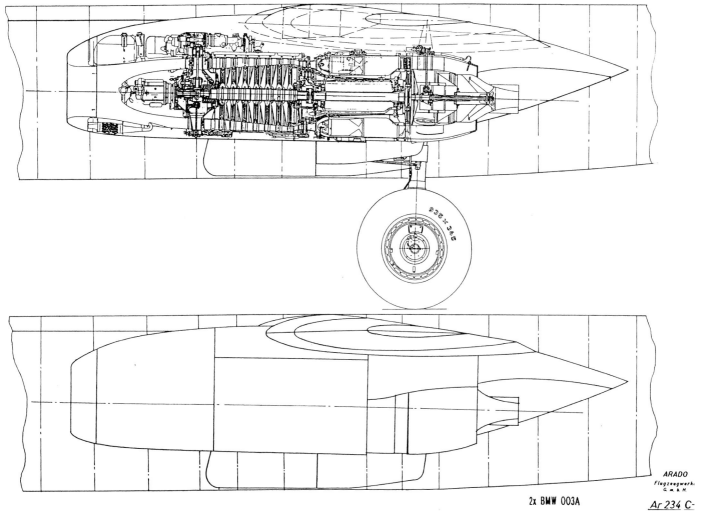

An *Arado* company pen and ink drawing (undated) of the outboard *BMW 003A-0* turbojet engine on a *Ar 234C-3*. Notice how *Arado* engineers considerably extended the wing's trailing edge out over the *engine's* exhaust nozzle as well as attempting to separate the thrust (for a moment) from its sister engine.

Arado Ar 234C

A view from beneath the port side twin *BMW 003A-0* turbojet engine nacelle. It appears that no bomb racks have yet been installed, however, the two large wing nuts seen in the bottom cowling area between the turbojets is to secure any bomb racks required as an effective "*Die Engländer Bomber.*"

A port side view of a *BMW 003A-0* turbojet engine with vertical fins on its thrust adjustable tail cone.

A ground level nose on view of the *Ar 234C-3's* tricycle landing gear. Scale model and photographed by *Günter Sengfelder*.

Arado Ar 234C

Wheels! This is the first *Ar 234C* with retractable wheels and it is the *V13*...the 3rd prototype shown here from its nose port side. The outer cowling on the engine's air intakes appear, from the photograph, to be painted a different color from the rest of the engine...perhaps red.

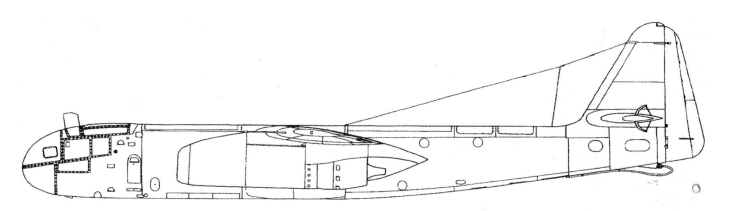

A pen and ink drawing from Arado featuring the port side profile of the *Ar 234 V13*.

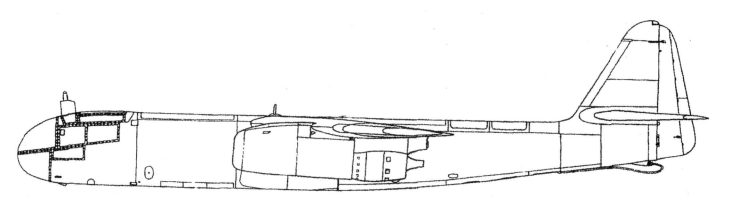

A pen and ink drawing from *Arado* featuring the port side of their proposed *Ar 234V16*, which was to be test flown with *Rüdiger Kosin's Versuchsflügel II* triple swept back wings.

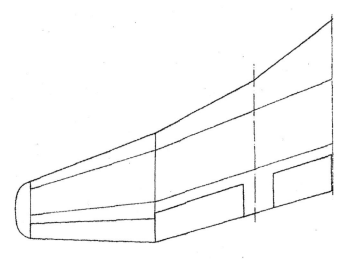

A pen and ink drawing of the triple swept back wing (Versuchsflügel II) proposed by *Arado's* chief aerodynamitist for the testing on the *Ar 234 V16*.

A full port side view of the *Ar 234 V13*, we*rk nummer 130023*, coded *PH+SU*, and powered by *4xBMW 003A-0* turbojet engines. Its maiden flight occurred on 6 September 1944, and the machine was destroyed on the same day. Notice that it does not appear to have provisions for a braking parachute.

A ground level nose on view of the *Ar 234 V19*. The periscope above the cockpit appears to be a similar to the *RF2C* sighting mechanism used on the *Ar 234B*. The *V19* is considered to be the first true *Ar 234C* prototype.

Arado Ar 234C

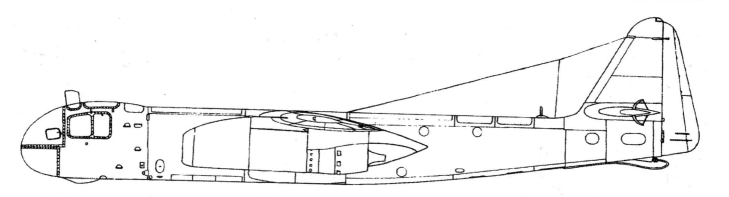

A pen and ink drawing from *Arado* featuring the port side of their *Ar 234 V19*.

A port side view of the *Ar 234 V19*. The surfaces appear painted in various places, however, the cowling around the port *BMW 003A-0* turbojet engines appear unpainted, as as well as its vertical stabilizer. Its maiden flight occurred on 16 October 1944. It was known to still exist as of April 1945.

The port side cockpit window pattern of an *Ar 234B*. Actually, the cockpit was pretty much glassed over. Visibility was greatly improved in the *Ar 234C* by making the side plexiglass into two individual windows.

Arado Ar 234C

The port side cockpit window pattern of an *Ar 234C*. Note the differences between it and the *Ar 234B*.

A nice nose port side view of the *Ar 234 V19*. This model appears to have a parachute breaking system installed. The *V19* was found by the Allies in April 1945, pretty much intact. It is not known what became of this machine postwar.

A new looking *Ar 234 V19* as seen at the tarmac's edge.

Ar 234 V20, werks nummer 130030, coded *PI+WY,* and powered by 4x*BMW 003A-1* turbojet engines. This was the first *Ar 234C* with full cockpit cabin pressurization. Its maiden flight occurred on 5 November 1944. Lost 4 April 1945 at Wesendorf due to an Allied bombing raid.

The *Ar 234C* as viewed from its starboard side and featuring the starboard engine nacelle with its 2x*BMW 003A-0* turbojet engines. Notice the position of the cockpit entry/exit hatch above the fuselage nose when it is fully open.

A ground level nose on view of the *Ar 234 V21.* This machine appears to be fully equipped, as well it should, because it was considered the final prototype for the *Ar 234C* series. Its maiden flight occurred on 24 November 1944. It was known to still exist as of February 1945.

Arado Ar 234C

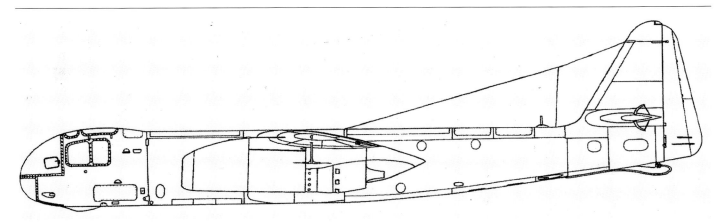

A pen and ink drawing from *Arado* featuring the port side of their *Ar 234 V21*.

The *Ar 234 V21* and featuring a wider and taller cockpit cabin. This machine was the pre-production version of the *Ar 234C-3* bomber version.

Ar 234 V21 as seen from its rear starboard side.

Arado Ar 234C

Ar 234 VERSUCHS-FLÜGEL II

PROFIL	1	2	3	4
WÖLBUNG %	1.25	1.25	1.20	1.20
WÖLBUNGSRÜCKLAGE %	25	25	25	25
PROFILDICKE %	09	09	10.5	10.5
NASENRADIUS · 1/d	1.16	1.16	1.027	1.027
DICKENRÜCKLAGE %	30	30	35	35

Ar 234 VERSUCHS-FLÜGEL III

PROFIL	1	2	3
WÖLBUNG %	1.0	1.0	1.0
WÖLBUNGSRÜCKLAGE %	50	50	50
PROFILDICKE %	09	11	12
NASENRADIUS · 1/d	0.33	0.36	0.375
DICKENRÜCKLAGE %	50	50	50

Ar 234 VERSUCHS-FLÜGEL IV

PROFIL	1	2	3	4
WÖLBUNG %	1.0	1.0	1.0	1.0
WÖLBUNGSRÜCKLAGE %	50	50	50	50
PROFILDICKE %	09	10	10.5	10.5
NASENRADIUS · 1/d	0.33	0.33	0.33	0.33
DICKENRÜCKLAGE %	50	50	50	50

Ar 234 VERSUCHS-FLÜGEL V

PROFIL	NACA 00012
WÖLBUNG	–
WÖLBUNGSRÜCKLAGE	–
PROFILDICKE	–
NASENRADIUS · 1/d	0.55
DICKENRÜCKLAGE %	50

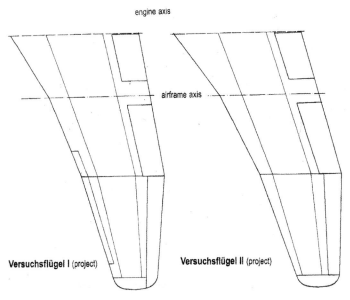

A pen and ink drawing from *Arado* featuring their *Versuchsflügel I* and *II* wing sweep back arrangements for upcoming *Ar 234Cs*. No *Arado 234B* or C model machines were completed with a swept back wing prior to war's end.

Arado's chief aerodynamist *Rüdiger Kosin* was in the process of testing four different wing shapes for *Walter Blume's Ar 234C*. *Kosin's* work was known as "*Ar 234 Versuchsflügel #1, #2, #3, and #4.*"

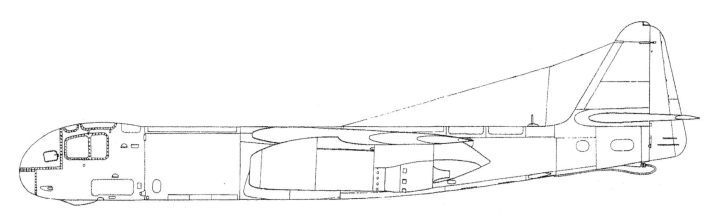

A pen and ink drawing from *Arado* featuring the starboard side of their *Ar 234 V21* with its proposed wing conversion to *Versuchsflügel IV* or a triple swept back wing.

Arado Ar 234C

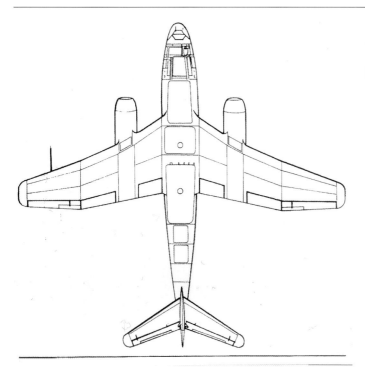

A fuselage bulkhead believed to be from the *Ar 234 V16* with its *Versuchsflügel I* swept back wing.

A pen and ink drawing of the experimental twin turbojet powered *Ar 234 V16* coded *PH+SX* which was to have featured the *Versuchsflügel I* or the crescent sweep back wing of *Arado* chief aerodynamitist *Dipl.-Ing. Rüdiger Kosin*. This machine, which had been assigned *werk nummer 130026*, was under construction at war's end. It's fuselage along with the *Versuchsflügel I* wing were later removed to England for investigation.

A wing tunnel model of a *Ar 234* which appears to be fitted with the *Versuchsflügel I* swept back wing.

Arado Ar 234C

In the foreground is a engineless *Ar 234C-3...werk nummer 250001*. The port wing of an unidentifiable *Ar 234C-3* looms over head.

Above: An *Ar 234C-3* thought to be the 6th pre-production example. It has been equipped with the shorter, angled, and more aerodynamic periscope.

Found post war in a scrap heap is this *Ar 234C-3 werk nummer 250012*. Notice the "*012*" stenciled on its port side vertical stabilizer.

Arado Ar 234C

Two American GI's shown standing on the port side of the *Ar 234C-8* bomber prototype. Notice how its *MG 151 20* mm cannon barely extends beyond the fuselage skin. Notice, too, that the cannon access hatch has been removed. It is not known what happen to *Ar 234C-8* at war's end. It was suppose to be equipped with 4x*Jumo 004D* turbojet engines. The *004D* was an improved version of the *004B* with a new fuel regulator which prevented too rapid throttle movement, and two-stage fuel injection. With the *004D* its full thrust of 2,200 pounds was permissible at all heights and speeds. It appears that no Ar 234C's were collected by the Allies. Numerous *Ar 234Bs*, however, were flight tested by several Allied powers post war but only one example exists in the world. It is located at NASM, Washington, D.C.

Arado engineers were uncertain where to place the 3rd and 4th *BMW 003A-1* turbojet engines on the *Ar 234B's* wing. So tests were done to determine if two turbojet engines in a single nacelle was better than four individually mounted units. This is a ground level nose on view of the *Ar 234 V8* with its four turbojet engines mounted in two nacelles.

Arado Ar 234C-1

Mission: Aerial Photo Reconnaissance

Mission-Carrying Equipment:
- 2x*Rb 50/30* or *Rb 75/30* photo cameras mounted inside rear fuselage forward the tailplane assembly;
- 2x66 gallon gasoline drop tanks...one under each dual engine nacelle;
- 1x*R2* under-fuselage cannon pod carrying 2x*MG 151 20 mm* [250 rounds per cannon] fixed cannon firing rearward...for self-defense;

Status:
- Never built. Project canceled May 1944. However, a photo reconnaissance version would be designed/built by modifying the basic *Ar 234C-3* flying machine.

A twin *MG 151* 20 mm cannon arrangement found in the *Rüstsätze* modification kit *R1* which would have been attached to the bottom fuselage pod of many *Ar 234C-3* variants.

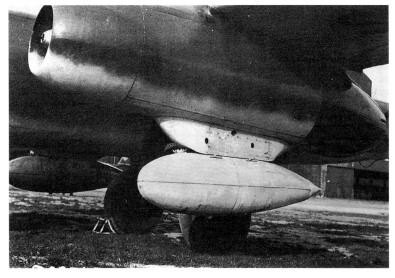

The port side view of the *Ar 234 V9* with an external 66 gallon fuel tank attached to its *ETC 504* bomb rack. This arrangement would be similar for the *Ar 234C* series and variants.

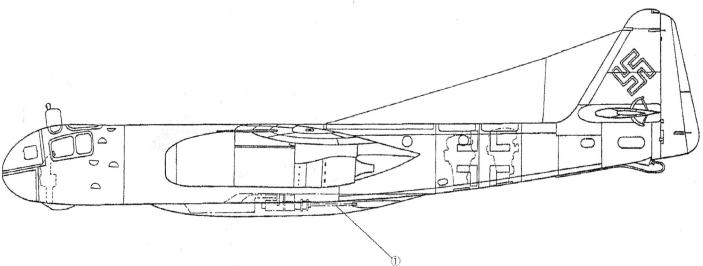

A pen and ink drawing by *Arado* featuring the starboard side of their proposed *Ar 234 C-1* reconnaissance machine.

Arado Ar 234C

A view of *Ar 234 C-2's* cockpit mockup. Shown is the machine's starboard side with the pilot's seatback visible in the lower right-hand corner.

A view of *Ar 234C-2's* cockpit mockup. Shown is the cockpit's rear port side. The pilot's seatback is visible in the bottom center of the photograph.

A view of *Ar 234C-2's* cockpit mockup. Shown is the starboard side of the cockpit. The pilot's seatback is seen in the lower right-hand corner of the photograph. Notice, too, that the yoke of the control column seen in this photo has a different shape and was not used again. Perhaps this photo is of the initial mockup with changes made as work progressed.

A view of *Ar 234C-2's* cockpit mockup. Shown is the cockpit's port side instruments. The pilot's seatback appears in the left-hand bottom corner of the photograph.

A view of *Ar 234C-2's* cockpit mockup. Seen is the starboard side of the cockpit with a full range of instruments installed for official viewing. The pilot's seatback can be seen in the lower right-hand corner of the photograph.

Arado Ar 234C

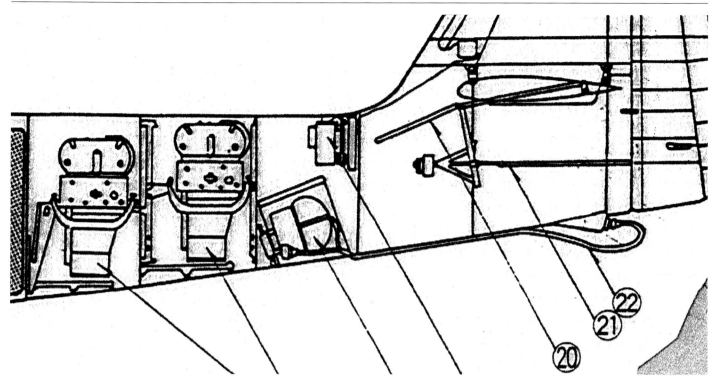

A port side close up view of the *Ar 234C-4's* twin *Rb 75/30* high resolution photo reconnaissance cameras mounted forward the brake-parachute box and vertical stabilizer.

The *Rb 50/30* aerial camera used by the *Luftwaffe* for making reconnaissance photographs. It had a focal length of 50 centimeters [19 1/2 inches] and an image size of 30 by 30 centimeters [11 3/4 by 11 3/4 inches].

The upper fuselage of an *Ar 234*, just forward of the vertical stabilizer, with its two access panels removed to show where each of the two photo reconnaissance cameras were located.

A close up of the upper aft fuselage of an *Ar 234* with its two camera hatches removed. It shows 2x*Rb 50/30* photo reconnaissance cameras. This pair of cameras were fitted with 50 centimeter telephoto lenses, the cameras pointed downwards, and were slanted sideways away from each other at 12% to the vertical across the line of flight. From an altitude of 32,500 feet [10,000 meters] this split pair camera arrangement took in a swathe of ground just over 6 miles [10 kilometers] wide along the *Ar 234's* flight track.

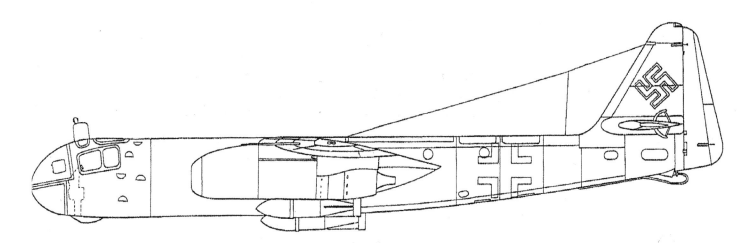

A pen and ink drawing by *Arado* of their *Ar 234C-2* bomber featuring its port side. The *C-2* is shown with *3xSC 500* kilogram [3x1,100 pound] bombs.

Arado Ar 234C-2

Mission: Aerial Bomber with 3,300 pound bomb load)

Mission-Carrying Items:
- 1x2,200 [*SC 1000* kilogram] bomb plus 1x1,100 [*SC 500* kilogram] bomb both under the center fuselage;
- 1x2,200 [*SC 1000* kilogram] bomb plus 2x550 [*SC 250* kilogram] bombs under the center section;
- 3x1,100 [*SC 500* kilogram] bombs: one under the center fuselage and one under each dual engine nacelle;
- 2x*MG 151 20* mm[250 rounds per cannon] fixed cannon firing rearward in rear fuselage...for self-defense;
- 2x*MG 151 20* mm [250 rounds per cannon] fixed cannon firing forward in the nose beneath the cockpit...for self-defense;

Status:
- Never built. Project canceled May 1944. However, another aerial bomber version would be designed/built by modifying the basic *Ar 234C-3* flying machine.

Common aerial bombs used by the Luftwaffe. Top to bottom to scale:
- *BM 1000-G* (2,205 lb/1,000 kg) mine
- *SD 1000* (2,205 lb/1,000 kg) armor piercing bomb
- *SD 500-II* (1,102 lb/500 kg) armor piercing bomb
- *SD 500-E* (1,102 lb/500 kg) armor piercing bomb
- *SD 500-A* (1,102 lb/500 kg) splinter bomb
- *SC 500* (1,102 lb/500 kg) bomb
- *SD 500* (1,102 lb/500 kg) bomb
- *SC 2000* (4,412 lb/2,000 kg) bomb
- *SC 1800* (3,968 lb/1,800 kg) bomb
- *SD 1400* (3,086 lb/1,400 kg) armor piercing bomb
- *SC 1000-L* (2,205 lb/1,000 kg) bomb
- *SC 1000* (2,205 lb/1,000 kg) bomb
- *SD 250* (551 lb/250 kg) splinter bomb
- *SC 250* (551 lb/250 kg) bomb
- *SD 50* (110 lb/50 kg) splinter bomb
- *SC 50* (110 lb/50 kg) bomb
- *SC 10* (22 lb/10 kg) bomb

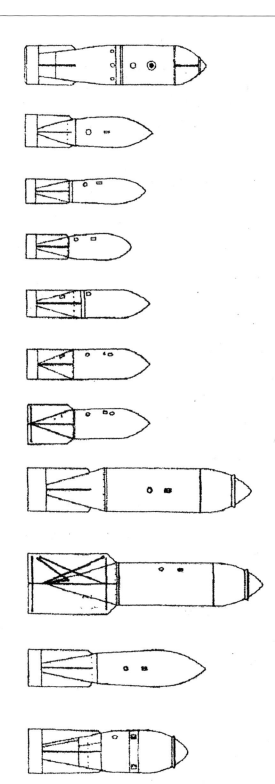
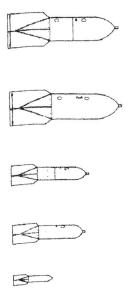

Arado Ar 234C-3

Mission: Basic Series Pattern *Ar 234C* Flying Machine. Prototype Also Known As The *Ar 234V21*.

Series Standard Equipment:
- 4x*BMW 003A-1* turbojet engines;
- 1x*PV1B* gunsight, 1x*BZA1C* bombsight, and 1x*Lofte 7D* bombsight;
- 1x*Schloss 2002* or *ETC 2000* bomb rack under each dual engine nacelle;
- Radio equipment: *FuG 16*, *FuG 25*, *FuG 102*, *FuG 136*, *FuG 142*, and *FuG 217*;
- 2x*MG 151 20* mm [250 rounds per cannon] fixed cannon firing rearward in rear fuselage...for self-defense;
- 2x*MG 151 20* mm [250 rounds per cannon] fixed cannon firing forward in the nose beneath the cockpit...for self-defense;

Status:
- First prototype of the *Ar 234C* series (*Ar 234 V19*) machine was first flown on 16 October 1944. Last flight occurred in April 1945. Blown up in the Hofoldinger Wood to keep from the advancing U.S. Army.
- Second prototype of the *Ar 234C* series (*Ar 234 V20*) was destroyed in a U.S.A.A.F. air raid on Wesendorf 4 April 1945.
- Third and final prototype of the *Ar 234C* series and series pattern (*Ar 234 V21*) first flew on 24 November 1944. It was later modified with swept-back *Versuchsflügel IV* wings. Modification work was not completed prior to war's end.

Below: Pen and ink drawing by *Arado* featuring their *Ar 234C-3* as viewed from its port side: #01 - 2x*MG 151 20* mm cannon with 250 rounds each; #02 - redesigned nose and cockpit; #03 - port side *MG 151* cannon access hatch; and #04 - loop antenna.

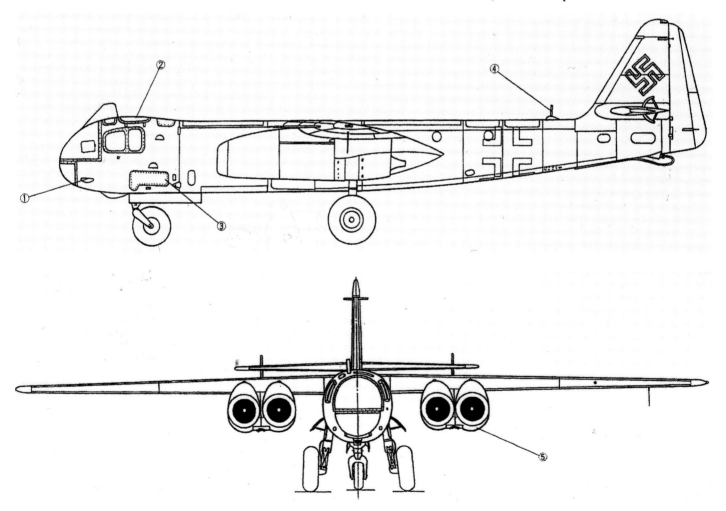

Pen and ink drawing by *Arado* featuring the front view of their *Ar 234C-3*: #05 - *BMW 003A-0* turbojet engine with its 1,764 pounds [800 kilograms] of thrust.

Arado Ar 234C

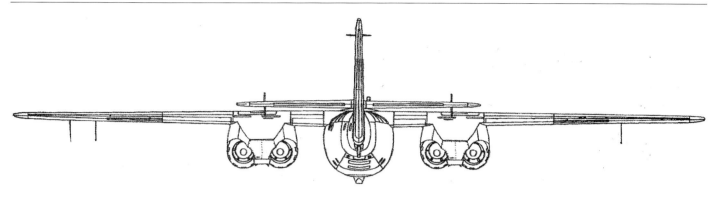

Pen and ink drawing by *Arado* featuring the rear view of their *Ar 234C-3*.

Front view of an *Ar 234C-3*. Scale model and photographed by *Günter Sengfelder*.

Port nose side, ground level view of an *Ar 234C-3*. Scale model and photographed by *Günter Sengfelder*.

Arado Ar 234C

Port nose, forward fuselage side, ground level view of an *Ar 234C-3*. Scale model and photographed by *Günter Sengfelder*.

Port nose, mid fuselage side, ground level view of an *Ar 234C-3*. Scale model and photographed by *Günter Sengfelder*.

Port side, ground level view of an *Ar 234C-3*. Scale model and photographed by *Günter Sengfelder*.

Arado Ar 234C

Port rear side, ground level view of an *Ar 234C-3*. Scale model and photographed by *Günter Sengfelder*.

Starboard front fuselage, ground level view of an *Ar 234C-3*. Scale model and photographed by *Günter Sengfelder*.

Starboard nose fuselage, ground level view of an *Ar 234C-3*. Scale model and photographed by *Günter Sengfelder*.

Arado Ar 234C

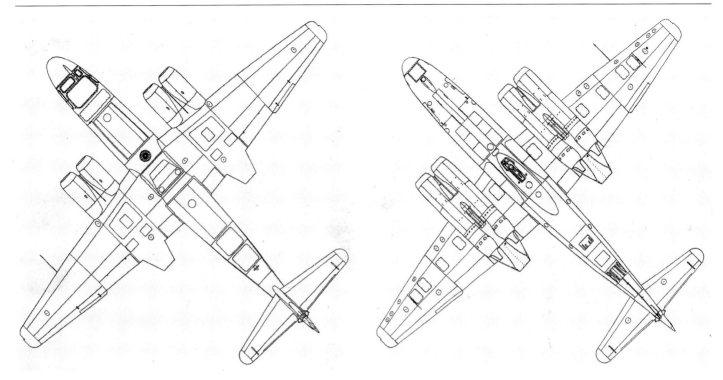

Pen and ink drawing by *Arado* featuring a top view of their *Ar 234C-3*.

Pen and ink drawing by *Arado* featuring a bottom view of their *Ar 234C-3*.

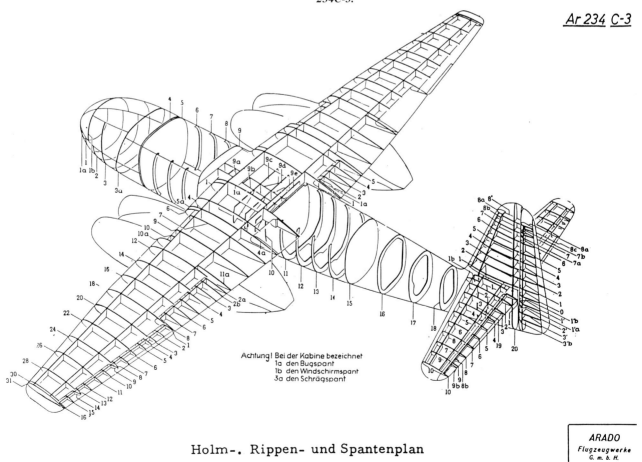

Holm-, Rippen- und Spantenplan

Achtung! Bei der Kabine bezeichnet
1a den Bugspant
1b den Windschirmspant
3a den Schrägspant

Ar 234 C-3

ARADO
Flugzeugwerke
G. m. b. H.

An *Arado* company pen and ink drawing featuring the *Ar 234C-3*'s fuselage bulkheads, wing spar, wing rib location, tail assembly spar and ribs. Date unknown, but probably ca 1944.

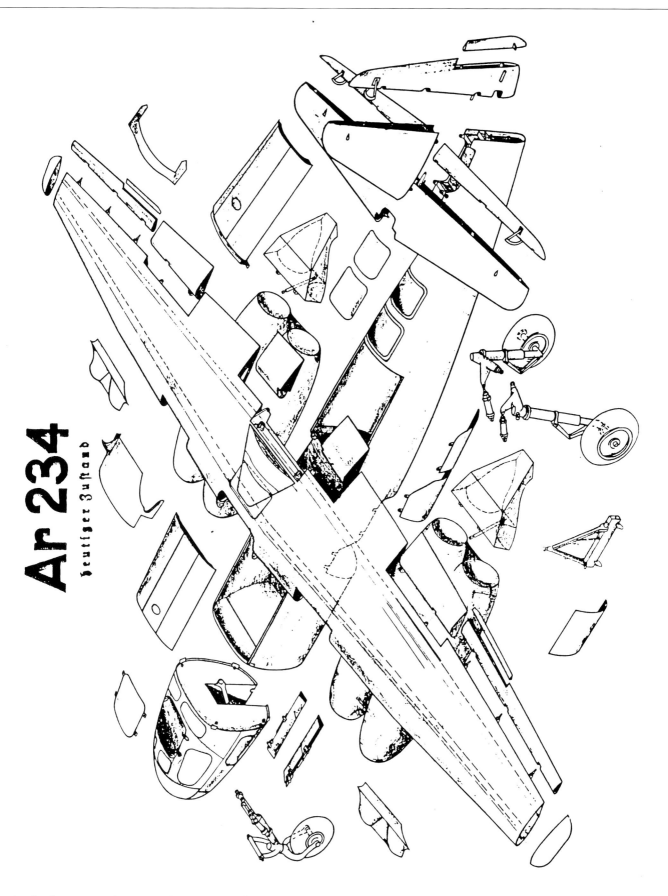

An *Arado* company document (date unknown) featuring the component parts making up their *Ar 234C-3*.

Arado Ar 234C

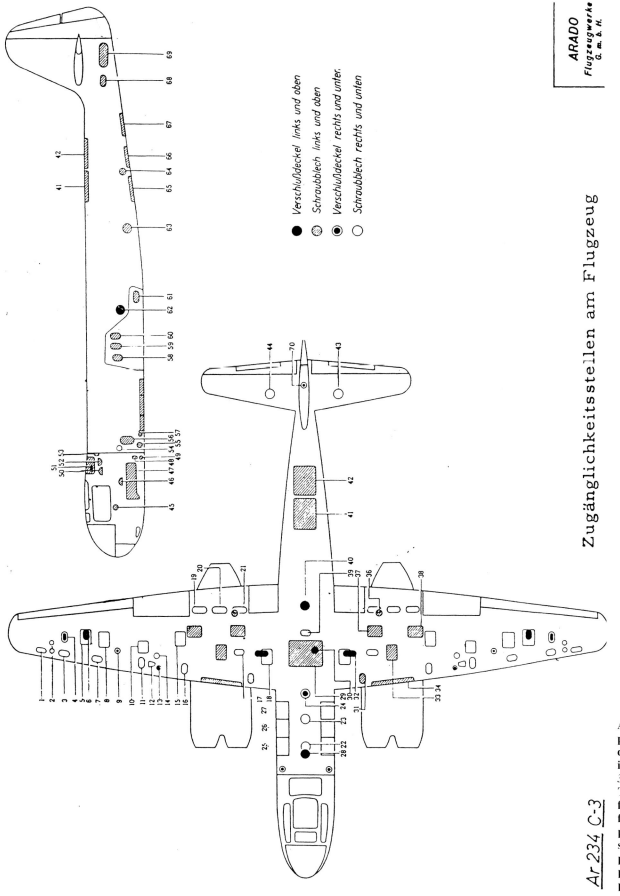

An *Arado* company pen and ink drawing featuring the *Ar 234C-3*. No date given. This drawing features all ports found on the upper wing and fuselage: access ports, inspection ports, fuel filler ports, and so on.

Arado Ar 234C

A poor quality pen and ink drawing by *Arado* featuring a port side top view of their *Ar 234C-3* in flight.

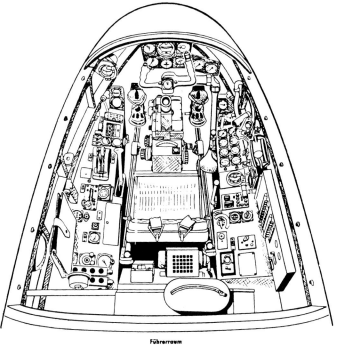

A pen and ink drawing by *Arado* featuring a general layout of the cockpit as found in the *Ar 234B-2*.

Left: A pen and ink drawing by *Arado* featuring the evolution of the cockpit cabins beginning with the *Ar 234A*, *234B-2*, *234C-3*, and *234C-5*.

Arado Ar 234C

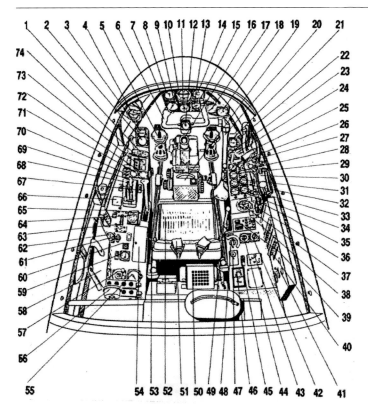

A pen and ink illustration of the open cockpit hatch on an *Ar 234B-2*. The single-seat *Ar 234C* looked pretty much the same. The throttles for the 2x*BMW 003A's* were located to the pilot's left, turbojet engine gauges located to the pilot's right, and flight instruments located in a panel behind the control column.

A wooden cockpit mockup for the single seat *Ar 234C-1*. Notice that the throttles for the 4x *BMW 003A* turbojet engines are on the pilot's left as they are in the *Ar 234B-2*. Flight instruments have been divided between the port and starboard side of the cockpit and not in the center in a panel behind the control column. Engine instrumentation is, again, located on the pilot's right. Located between the pilot's feet is a large box for the *Lotfe 7D* bomb aiming sight.

*Ar 234C-3 cockp*it cabin mockup by *Arado* featuring its starboard side.

Port side cockpit interior wooden mockup of the single-seat *Ar 234C-1*. Notice the four throttle levers (located in the middle of the photograph) for the machine's 4x*BMW 003A* turbojet engines.

Arado Ar 234C

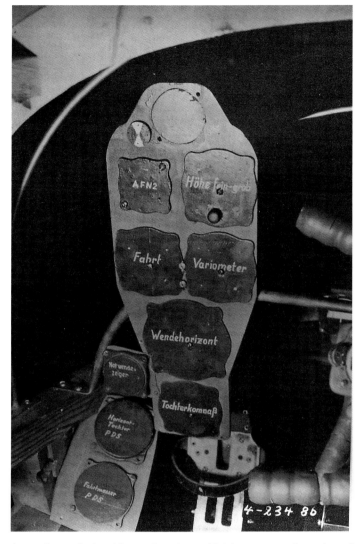

A front view of the *Ar 234C-3* cockpit cabin mockup by *Arado*.

A wooden cockpit cabin mockup (port side) instrument cluster/panel in the single seat Ar 234C-1. It reads top to bottom:
- *AFN2* - homing display
- *Höhe* - altimeter
- *Fahrt* - air speed indicator
- *Variometer* - meters per second
- *Wendehorizont* - artificial horizon
- *Tochterkommpass* - compass (the instrument panel)

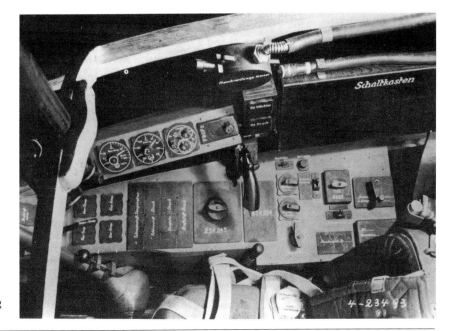

A wooden mockup of the instruments located along the *Ar 234C-3*'s starboard side cockpit by *Arado*.

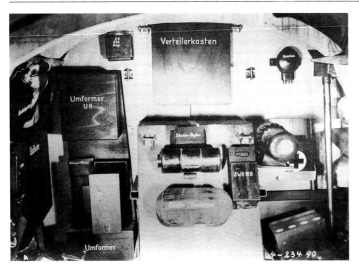

Aft cockpit cabin bulkhead in the wooden mockup of the *Ar 234C-3*'s cockpit and featuring transformers, emergency medical supplies, and so on.

Rear cockpit cabin bulkhead as shown here on *Arado*'s Ar 234C-3 wooden mockup featuring the 4x*BMW 003A-0* throttles seen in the lower right-hand corner of the photograph. Notice the flare gun's handle in the lower left-hand corner of the photograph.

Forward view of the *Ar 234C-2*'s wooden cockpit cabin mockup.

Forward view of the fully equipped *Ar 234C-3* wooden cockpit cabin mockup by *Arado*. Numerous changes were made in this version compared to the *Ar 234C-2*'s wooden mockup.

Arado Ar 234C

A view looking out of the *Ar 234C-3*'s wooden cockpit cabin mockup.

Wooden cockpit cabin mockup of the *Ar 234C-3* featuring the pilot's control column.

Inside the wooden cockpit cabin mockup for the proposed *Ar 234C-3*. Featured is the port side of the cockpit. Instruments seen include: gyro compass indicator, rate of climb indicator, and artificial horizon.

Front view of a version of the *Ar 234C-3*'s wooden cockpit cabin mockup.

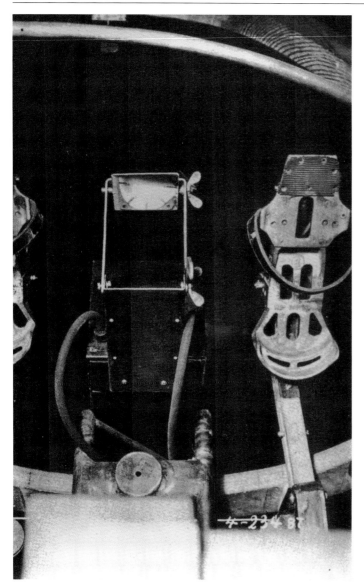

A wooden mockup of the "Lotfe D" bomb aiming sight installed in the Ar 234C's cockpit. The white painted "Lotfe D" is pointing forward.

Wooden cockpit cabin mockup of the Ar 234C-3 featuring its FuG 102 radio instrument located between the pilot's foot rudder pedels.

A view of the inside starboard side of a wooden cockpit cabin mockup for the Ar 234C-3.

Arado Ar 234C

Starboard side view of a wooden cockpit cabin mockup for an *Ar 234C-3* showing the proposed instrument layout along the starboard fuselage wall.

A view of the inside nose starboard side of a wooden cockpit cabin mockup for the *Ar 234C-3*. To the far right can be seen the right side rudder pedal.

Port side view of a wooden cockpit cabin mockup for an *Ar 234C-3*. Notice the 4x*BMW 003A-0* engine throttles on the console near the fuselage wall. The pilot's back seat in seen in the lower part of the photograph.

An overhead view of an *Ar 234C-3* featuring its control surfaces especially the ailerons and landing flaps on its wing's trailing edge. Scale model and photographed by *Günter Sengfelder*.

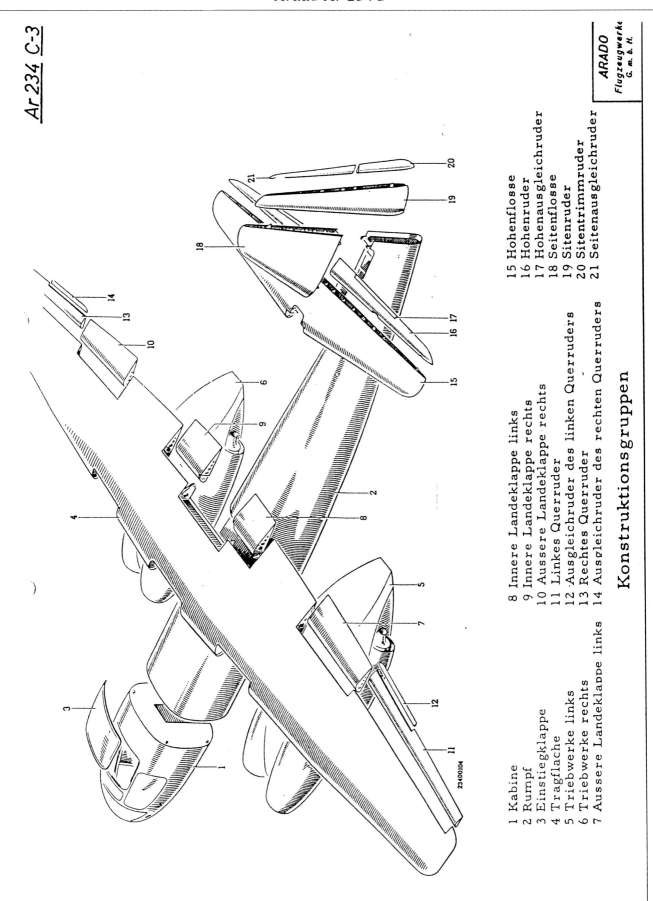

Ar 234 C-3

1 Kabine
2 Rumpf
3 Einstiegklappe
4 Tragflache
5 Triebwerke links
6 Triebwerke rechts
7 Aussere Landeklappe links
8 Innere Landeklappe links
9 Innere Landeklappe rechts
10 Aussere Landeklappe rechts
11 Linkes Querruder
12 Ausgleichruder des linken Querruders
13 Rechtes Querruder
14 Ausgleichruder des rechten Querruders
15 Hohenflosse
16 Hohenruder
17 Hohenausgleichruder
18 Seitenflosse
19 Sitenruder
20 Sitentrimmruder
21 Seitenausgleichruder

Konstruktionsgruppen

ARADO
Flugzeugwerke
G. m. b. H.

Arado Ar 234C

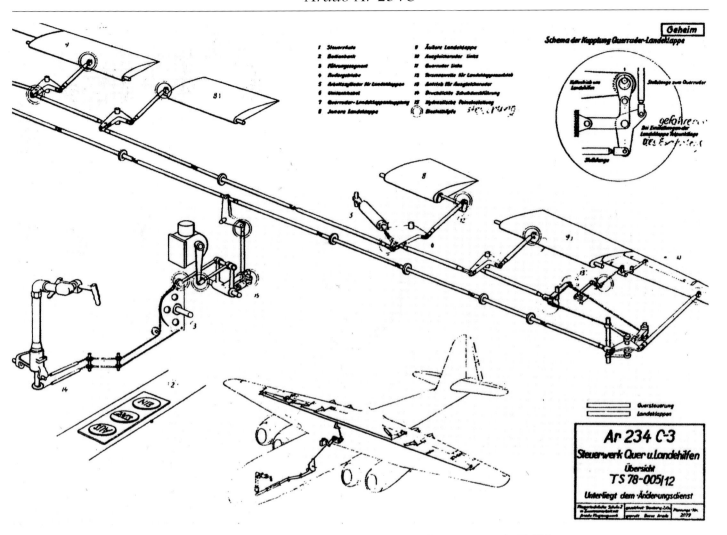

A pen and ink drawing by *Arado* featuring the entire aileron and landing flap control system on an *Ar 234*C-3.

Opposite: An *Arado* company document (undated) featuring the various control surfaces of the wings and tail assembly for their *Ar 234C-3*:

01 - Nose/cockpit.
02 - Fuselage.
03 - Cockpit entrance/exit hatch.
04 - Wings.
05 - Port (left) dual *BMW 003A* turbojet engines.
06 - Starboard (right) dual *BMW 003A* turbojet engines.
07 - Port outboard landing flap.
08 - Port inboard landing flap.
09 - Starboard inboard landing flap.
10 - Starboard outboard landing flap.
11 - Port aileron.
12 - Port aileron trim tab.
13 - Starboard aileron.
14 - Starboard aileron trim tab.
15 - Port aileron leading edge cover.
16 - Port horizontal stabilizer.
17 - Port elevator.
18 - Port elevator trim tab.
19 - Starboard elevator.
20 - Starboard elevator trim tab.
21 - Vertical stabilizer.
22 - Hinged rudder.
23 - Upper rudder trim tab.
24 - Lower rudder trim tab.

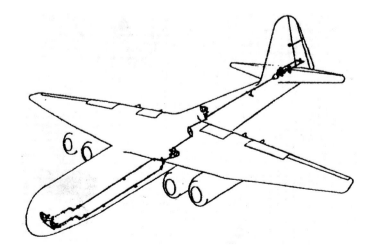

A poor quality pen and ink drawing by *Arado* featuring the control linkage path between the cockpit and the *Ar 234C-3*'s vertical hinged rudder.

An *Ar 234C-3* showing its four individual landing flaps in the full down position. Scale model and photographed by *Günter Sengfelder*.

An *Ar 234C-3* as seen from above with its landing gear extended and its landing flaps in the full down position for landing. Scale model and photographed by *Günter Sengfelder*.

A ground level rear port view of an *Ar 234C-3* and featuring its landing flaps in the full down position. Scale model and photographed by *Günter Sengfelder*.

Arado Ar 234C

The *Ar 234C-3/006* (6th pre-production version) seen from ground level with its landing flaps full down.

The tail assembly of an *Ar 234C-3* as seen from its rear port side and featuring its vertical stabilizer with hinged rudder, and rudder trim tab. This rudder assembly was pretty the same as found on the *Ar 234B* with twin their *BMW 003A-0* turbojet engines. Scale model and photographed by *Günter Sengfelder*.

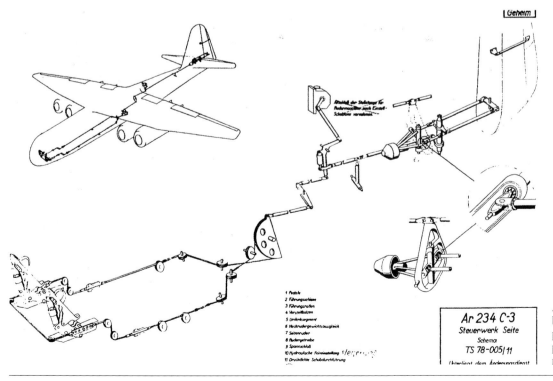

A detailed pen and ink drawing by *Arado* featuring the main points of the control linkage path to actuate the *Ar 234C-3's* rudder.

Arado Ar 234C

The tail assembly of an *Ar 234C-3* as seen from its port side and featuring its vertical rudder and upper and lower trim tabs. Scale model and photographed by *Günter Sengfelder*.

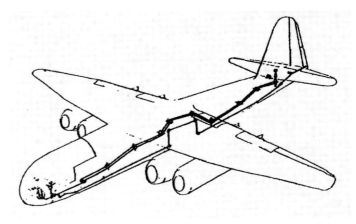

A poor quality general arrangement pen and ink drawing of the *Ar 234C-3's* rudder trim tab linkage between the cockpit and rudder.

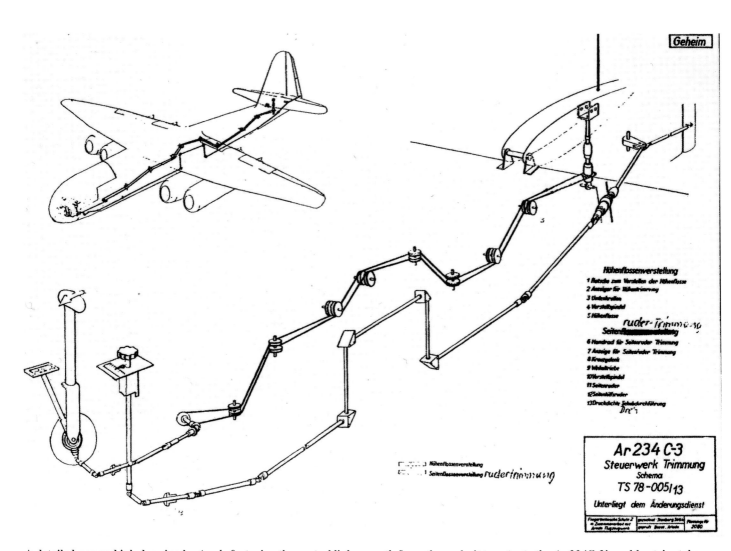

A detailed pen and ink drawing by *Arado* featuring the control linkage path from the cockpit to actuate the *Ar 234C-3's* rudder trim tabs.

Arado Ar 234C

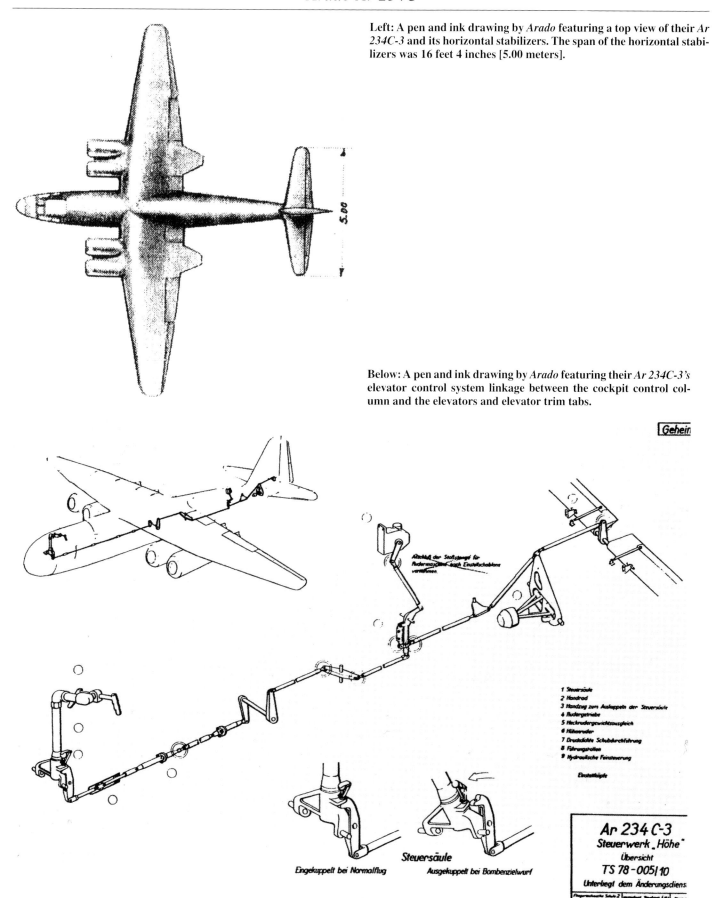

Left: A pen and ink drawing by *Arado* featuring a top view of their *Ar 234C-3* and its horizontal stabilizers. The span of the horizontal stabilizers was 16 feet 4 inches [5.00 meters].

Below: A pen and ink drawing by *Arado* featuring their *Ar 234C-3's* elevator control system linkage between the cockpit control column and the elevators and elevator trim tabs.

Arado Ar 234C

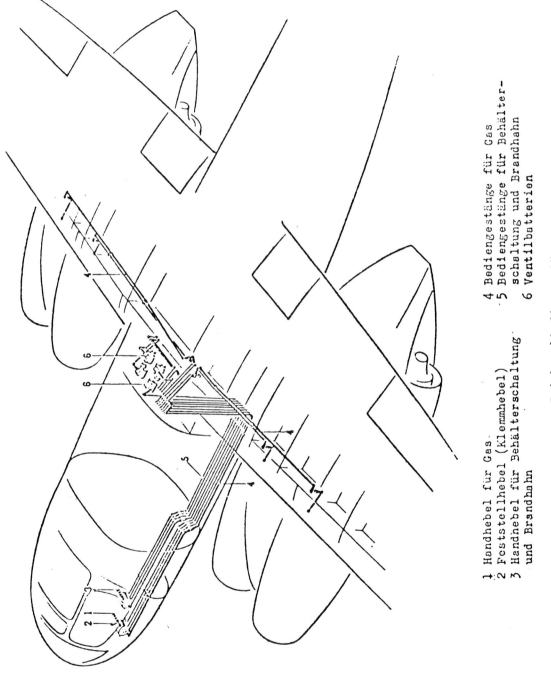

Ar 234 C-3

A pen and ink drawing from *Arado* featuring engine throttle linkage from the cockpit to 4x*BMW 003A-0* turbojet engines...a general arrangement in the *Ar 234C-3*.

Arado Ar 234C

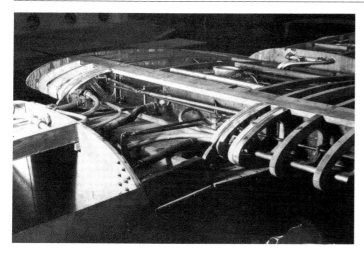

Wooden mockup of the *Ar 234C's* port side aft cockpit cabin and wing's leading edge. Aft the cockpit can be seen the route the hand throttle linkage takes in its journey to each of the four turbojet engines. Aft the wing's trailing edge in the upper fuselage are the machine's *J2* turbojet engine fuel tanks.

Wooden mockup of the *Ar 234C's* starboard side aft cockpit cabin and wing's leading edge. Aft the wing's trailing edge in the upper fuselage can be seen the machine's *J2* fuel tanks.

A wooden mockup of the *Ar 234A/B* and showing the *J2* turbojet engine fuel tanks aft the wing's trailing edge. Although this is a mockup of the *Ar 234A/B*, it would also apply to the *Ar 234C* series and variations.

A wooden mockup of the *Ar 234A/B* showing the *J2* turbojet engine fuel tanks looking forward to the cockpit. At the bottom of the photo is the wing's leading edge along with fuel pipes and linkage to the engine's throttle mechanism. Although this photo features a *Ar 234A/B*, it would also apply to the *Ar 234C* series and variations.

Arado Ar 234C

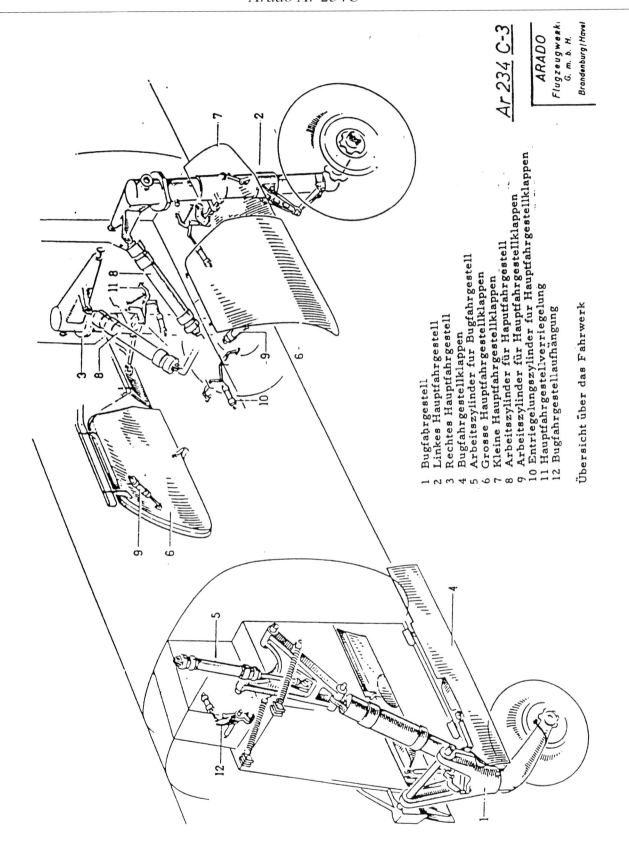

Ar 234 C-3

ARADO
Flugzeugwerk
G. m. b. H.
Brandenburg/Havel

1 Bugfahrgestell
2 Linkes Hauptfahrgestell
3 Rechtes Hauptfahrgestell
4 Bugfahrgestellklappen
5 Arbeitszylinder für Bugfahrgestellklappen
6 Grosse Hauptfahrgestellklappen
7 Kleine Hauptfahrgestellklappen
8 Arbeitszylinder für Hauptfahrgestellklappen
9 Arbeitszylinder für Hauptfahrgestellklappen
10 Entriegelungszylinder für Hauptfahrgestellklappen
11 Hauptfahrgestellverriegelung
12 Bugfahrgestellaufhängung

Übersicht über das Fahrwerk

Arado Ar 234C

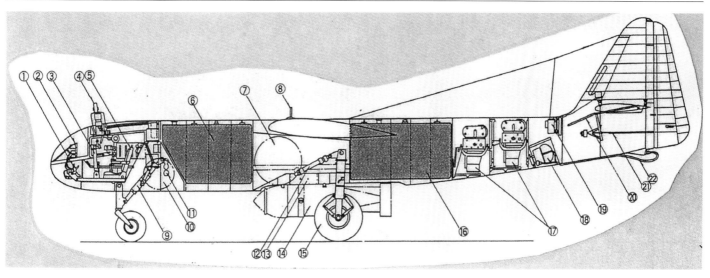

An *Arado* document featuring a *Ar 234B-1* photo reconnaissance machine which is also equipped to carry one *SC 1000* bomb externally under its fuselage.

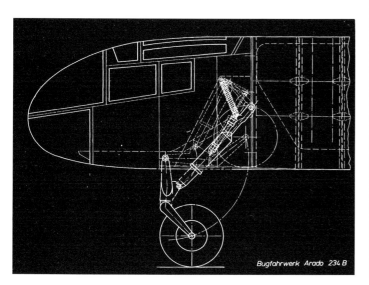

A pen and ink drawing of the *Ar 234B* series nose wheel arrangement as viewed from its port side. It featured two, side-by-side retraction cylinders. The Ar 234C series used only one large hydraulic retraction cylinder. Pen and ink drawing by *Günter Sengfelder*.

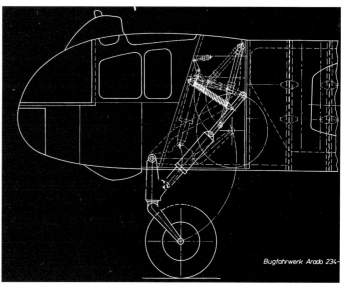

A pen and ink drawing of the *Ar 234C* series nose wheel arrangement as viewed from its port side. The "C" series had a revised, lengthened nose wheel fork, and a larger wheel brake than found in the "B" series. Pen and ink drawing by *Günter Sengfelder*.

Opposite: An *Arado* company pen and ink illustration (undated) featuring the tricycle landing gear's general arrangement on the *C-3*.
#01 - nose wheel support fork;
#02 - port main wheel oleo support strut;
#03 - starboard main wheel oleo support strut;
#04 - nose wheel gear cover door;
#05 - hydraulic cylinder for retracting nose wheel gear;
#06 - main wheel gear cover door, forward;
#07 - small main wheel gear, aft;
#08 - hydraulic cylinders - retracting main wheel gear;
#09 - hydraulic cylinders - retracting forward gear cover door;
#10 - brake fluid lines for main wheel brakes;
#11 - retracting extension for main wheel gear cover;
#12 - locking mechanism for nose wheel gear while retracted;

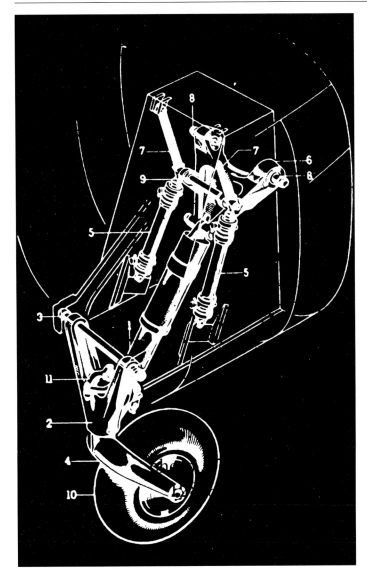

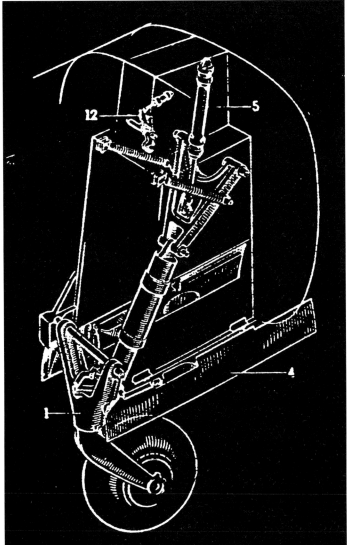

A pen and ink drawing of the nose landing gear on the *Ar 234B*. It featured twin hydraulic retraction cylinders. Pen and ink drawing by *Günter Sengfelder*.

Top Right: A pen and ink drawing of the nose landing gear on the *Ar 234C*. Changes were made from its sister *Ar 234B*. In addition to a larger nose fork, the landing gear on the *Ar 234C* used only one hydraulic retraction cylinder. It also had a larger wheel brake. Pen and ink drawing by *Günter Sengfelder*.

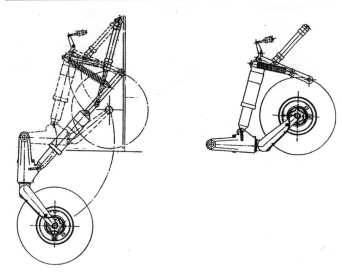

A pen and ink illustration featuring the nose landing gear on the *Ar 234C* series machines. Notice that upon retracting, the hydraulic retraction cylinder completely compresses in order to fit into its fuselage wheel well. Pen and ink drawing by *Günter Sengfelder*.

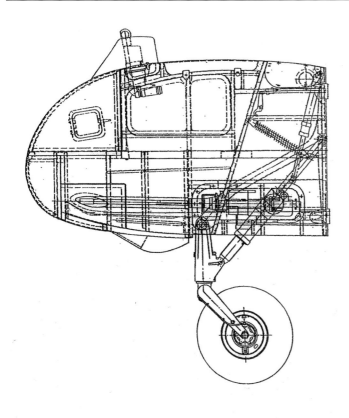

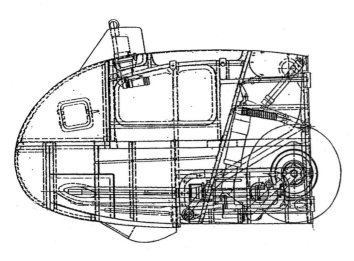

The nose landing gear on the *Ar 234C* series machines featuring the gear in fully down and locked position (top). The (bottom) illustration shows the nose landing gear fully retracted. Pen and ink drawing by *Günter Sengfelder*.

A wood and metal mockup of the *Ar 234B/C* featuring its nose landing gear and the hydraulic retraction cylinder looking forward. The port side nose gear cover is seen in the left middle of the photograph.

Arado Ar 234C

A wood and metal mockup of the *Ar 234B/C* featuring a good view of the machine's nose landing gear, metal gear door covers, hydraulic retraction cylinder, and nose wheel.

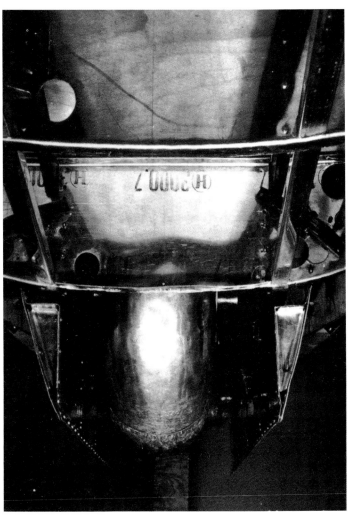

A wood and metal mockup of the *Ar 234B/C* featuring the machine's nose landing gear. In this photo we see the nose wheel is an almost full retracted position, however, the nose gear metal door covers are still in their open position.

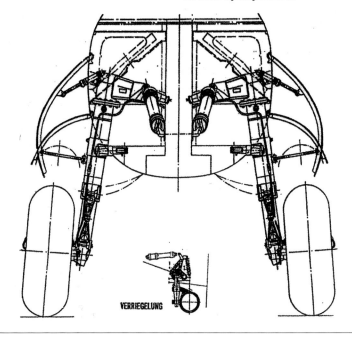

A pen and ink drawing featuring the fuselage section containing the main landing gear found on the *Ar 234B/C*. Looking aft. Pen and ink drawing by *Günter Sengfelder*.

Arado Ar 234C

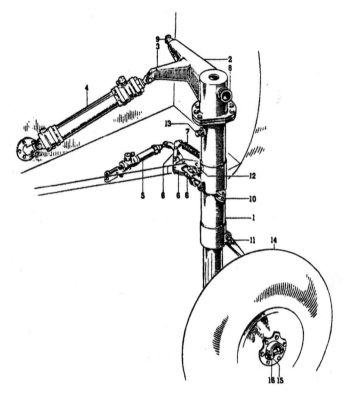

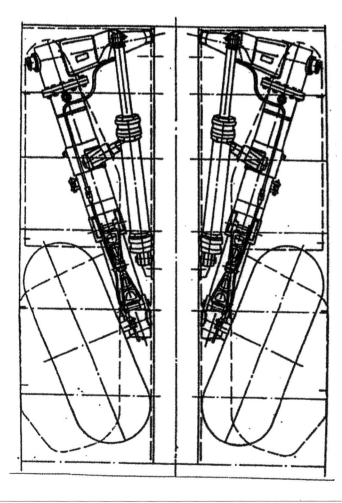

A wood and metal mockup of the *Ar 234B/C* as viewed from beneath the fuselage nose. Notice that the nose landing gear is fully retracted and its metal gear door covers are closed.

Above Right: The port side main landing gear on the Ar 234B/C. Items include:
#01 - shock absorber leg;
#02 - swivel leg;
#03 - swivel axis lever;
#04 - main landing gear hydraulic retraction cylinder;
#05 - main landing gear hydraulic locking cylinder;
#06 - locking mechanism;
#07 - pressure spring;
#08 - lynch pin;
#09 - lynch pin lock fitting for bayonet locking;
#10 - landing gear door attachment point;
#11 - steering control arm;
#12 - grease nipple;
#13 - filler screw;
#14 - main wheel tire size: 935x345 mm;
#15 - main heel axle;
#16 - axle cap;
Pen and ink drawing by *Günter Sengfelder*.

Right: A pen and ink drawing of the main landing gear on a Ar 234B/C when fully retracted. Seen from upper fuselage looking down. Pen and ink drawing by *Günter Sengfelder*.

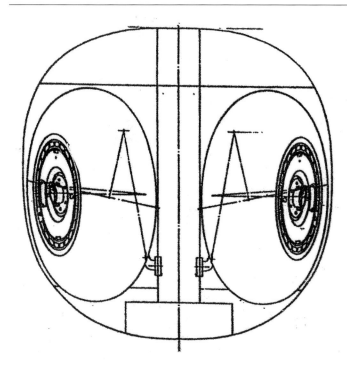

A pen and ink drawing of the main landing gear on an *Ar 234B/C* when fully retracted as viewed when looking aft. Pen and ink drawing by *Günter Sengfelder*.

A full size wood and metal mockup of the *Ar 234B/C* and featuring its port side main landing gear, metal gear door cover, metal sheet covered wooden main wheel, as well as a full scale bomb made out of wood. The mockup shown is seen looking aft toward the tail assembly.

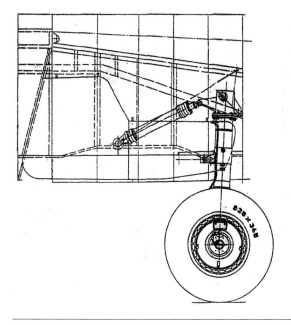

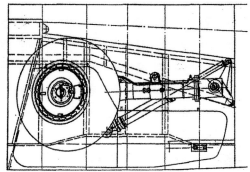

A pen and ink drawing featuring the position of the starboard side main landing gear wheel of an *Ar 234B/C* when extended (left) and retracted (right). Pen and ink drawing by *Günter Sengfelder*.

Arado Ar 234C

A wood and metal full-scale mockup of the *Ar 234B/C*. Its port side main landing gear appears to be fully retracted into the fuselage. The *Ar 234* had a two part gear door cover: an aft squarish door and a forward semi-circle-shaped door. Although it is difficult to see, the 2nd or squarish gear door is closing first and then next comes the forward semi-circle-shaped door.

A wood and metal full-scale mockup of the *Ar 234B/C* and featuring its port side main landing gear as viewed from ground level and behind.

A wood and metal full-scale mockup of the *Ar 234B/C* and featuring its port side main landing gear as viewed from its side and front. The wheel is in the process of retracting up into the fuselage. Notice the carved wooden *SC 1000* bomb under the fuselage between the two main wheels.

A wood and metal full-scale mockup of the *Ar 234B/C*. This photo of its port side was taken several feet away from the fuselage and wing root. The port side main gear is fully retracted and the two gear door covers have closed tight. The cockpit area is to the left in the photo while the tail assembly is to the right.

Arado Ar 234C

A close-up of the entry/egress hatch on the *Ar 234B/C*. This hatch covered opened to starboard and was held open by the metal rod seen to the right in the photograph.

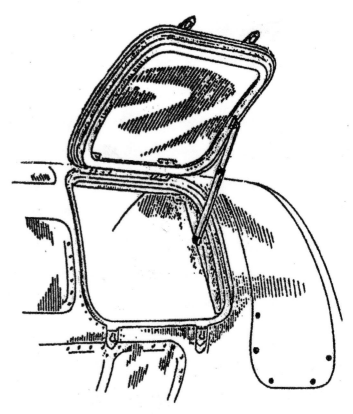

An *Arado* company pen and ink drawing of the open pilot/crew's entry and exit hatch on an *Ar 234C*. It opened to starboard and was held open by a hinged support bracket.

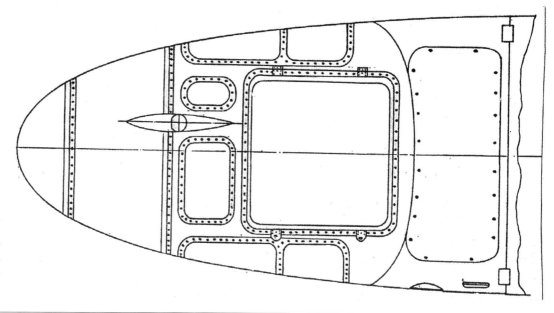

An *Arado* company pen and ink drawing of the *Ar 234C's* upper cockpit roof. The large square block is the pilot/crew's cockpit entry and exit hatch. This hatch was hinged and opened starboard.

Arado Ar 234C

An *Ar 234C-3* as seen from its port side and ground level. Notice that its pilot/crew entry/exit hatch is open. Scale model and photographed by *Günter Sengfelder*.

A full open pilot entry and exit hatch on an *Ar 234B*. The pilot sitting on the fuselage before the open hatch would place both of his feet into the open cockpit supporting himself with both hands straddling the open hatch. Once this was accomplished, the pilot would slowly settle into the pilot's seat. The same approach was used for entering into the open hatch on the *Ar 234C* and both pilots complained that these machines were difficult to get in and out of and perhaps nearly impossible to bail out in an emergency.

Wolfgang Zeise atop his *Ar 234B-2* aft the open entry/exit hatch as it is being readied for a high altitude photo reconnaissance flight.

A member of *Zeise's* black-clothed ground crew assists him through the open hatch.

Zeise's ground crewman is ready to close and secure-down the entry hatch.

The incline of the leading edge of the periscope installed on the *Ar 234C-3* can be judged from this photo. Scale model and photographed by *Günter Sengfelder*.

Right: The cockpit periscope on an *Ar 234B*.

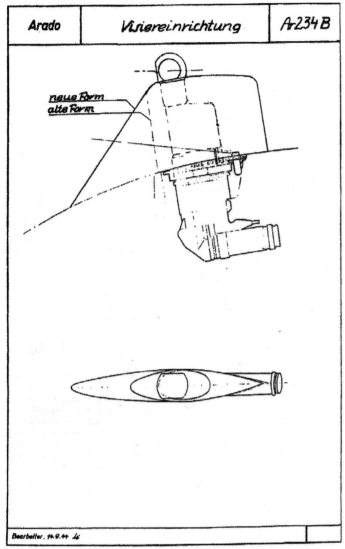 

A pen and ink drawing from *Arado* featuring the periscope installed on the *Ar 234C* series machines. The forward edge was more inclined and hence, more streamlined for less air resistance than the periscope installed on the *Ar 234Bs*.

The leading edge, more like that of a straight edge, of the *Ar 234B's* periscope.

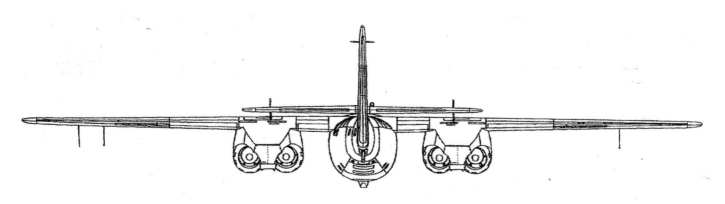

The several under wing antenna masks on the *Ar 234C-3s*.

Arado Ar 234C

The FuG 25a radio antenna mounted on the underside of the *Ar 234C's* wing.

A close-up view of the *FuG 25a IFF* (Identification: Friend or Foe) rod antenna.

A close-up view of the underside of the turbojet engine cowling on an *Ar 234C* where bomb racks could be quickly attached. The tear-drop concaved area aft the main landing gear for to provide sufficient space to carry one *Ar E.381* interceptor or one *SC 2000* bomb. Scale model and photographed by *Günter Sengfelder*.

An *Ar 234C-3* as seen from below and off to starboard with its landing flaps extended. On its port wing leading edge can be seen its pitot tube. Scale model and photographed by *Günter Sengfelder*.

Arado Ar 234C

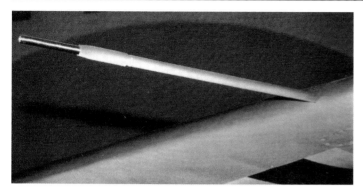

The complete pitot tube on an *Ar 234C-3's* port wing.

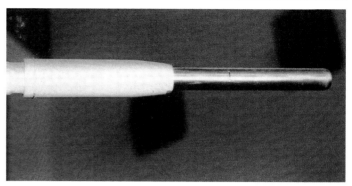

A close up view of the *Ar 234C-3's* pitot's leading edge.

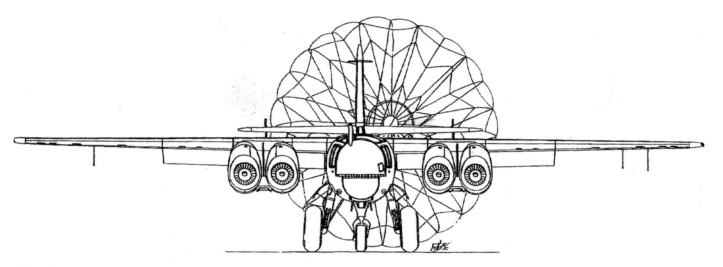

This is how the *Ar 234C-3's* and its variants might have looked when its braking parachute was fully deployed. Courtesy: *Marek Rys, Ar 234 Blitz, AJ Press.*

The *Ar 234C* required a parachute for breaking upon touch down like its sister *Ar 234B*. In both versions, the braking parachute was held in a compartment in the lower fuselage. The parachute was connected to a steel cable which was attached to a mounting beneath the vertical stabilizer. This steel cable can be seen beneath the *Ar 234C's* tail assembly. Scale model and photographed by *Günter Sengfelder*.

Arado Ar 234C

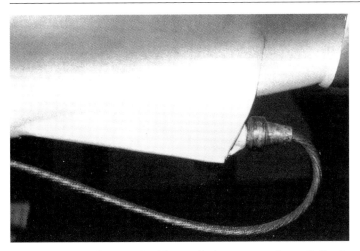

A close up view of the steel cable attached to a mount beneath the *Ar 234C's* tail assembly.

The braking parachute steel cable is secured to the *Ar 234C* immediately beneath its hinged rudder as shown in this photograph.

The braking parachute steel cable enters the parachute box in a *Ar 234C* under the aft fuselage.

The braking parachute steel cable shown entering into the parachute box on an *Ar 234C*.

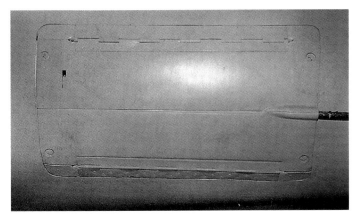

The braking parachute holding box in the *Ar 234C*. This box is under the fuselage and covered by two hinged doors. Upon release by the pilot after touch down, these doors open and the parachute comes out and deploys.

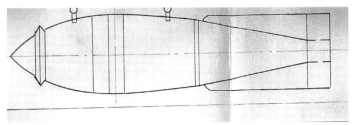

A typical *Luftwaffe* 2,205 pound *SC 1000* [kilometer] bomb.

Opposite: An *Arado* company pen and ink drawing (undated) featuring their *Ar 234C-3* and where 3x*SC 2000* kilogram (4,409 pound) bombs would be carried externally. With a fleet of *SC 2000* bomb-carrying *Ar 234C-3*s, *Adolf Hitler* could, at last, have his long awaited "*Engländer Bomber*."

Arado Ar 234C

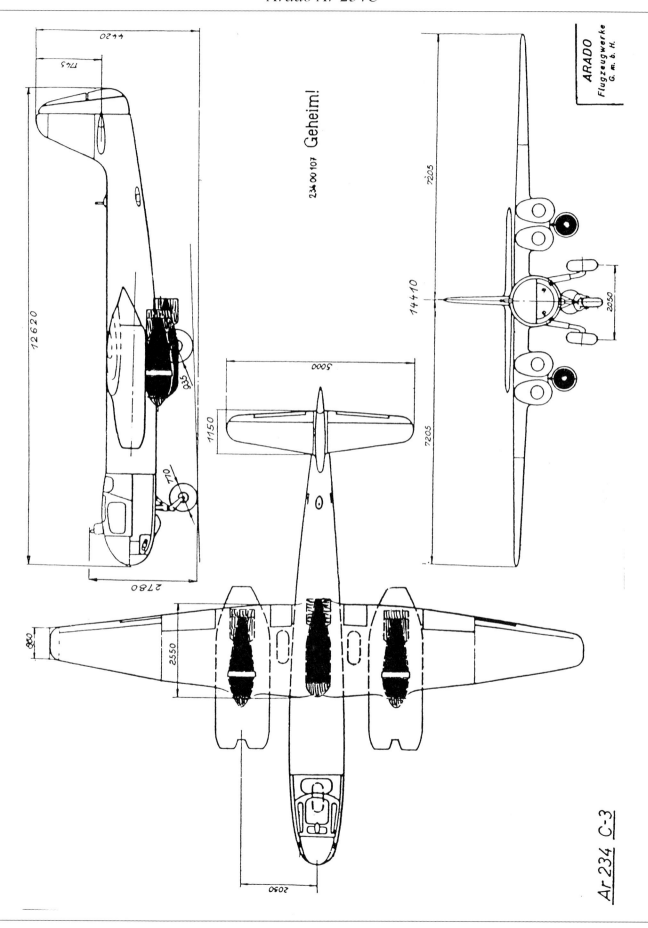

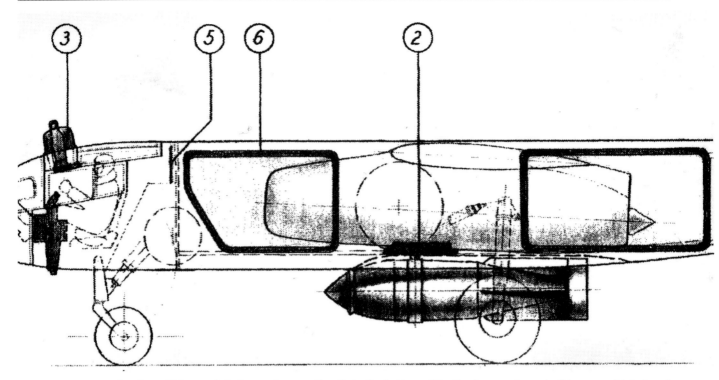

The center *SC 1000 bomb* would be carried via attachments directly to the bottom of the fuselage as shown in this pen and ink drawing.

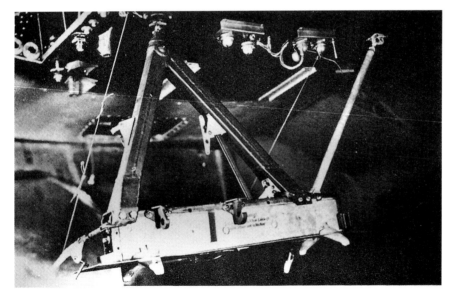

On the other hand, the *Ar 234C-3* required a special bracket configuration to hold an *SC 1000* bomb beneath its two turbojet engine nacelles. *Arado* engineers developed this frame so that the upper part bolted to the main spar. The bomb attachment metal frame hung down between the two *BMW 003A-0s* without touching either engine nacelle.

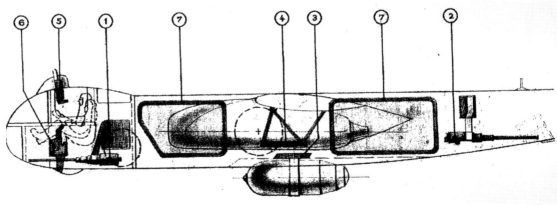

A pen and ink drawing featuring the special bomb attachment bracket which hung down between the *Ar 234C-3's BMW 003A-0* turbojet engines. See #3 and #4.

Arado Ar 234C

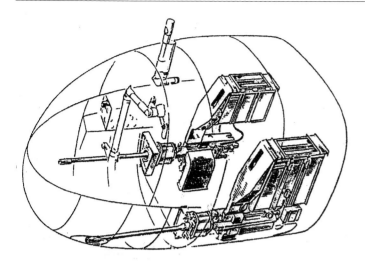

This is an *MG 151* 20 mm cannon of the type installed in most *Ar 234s*, including the *Ar 234C* series and its variations.

The *Ar 234C* series could be heavily armed depending on its use. Typically it was equipped with 2x*MG 151 20* mm cannon in the aircraft's nose beneath the cockpit. This pen and ink drawing features an *Ar 234* equipped with 2x*MG 151's* and ammunition storage box.

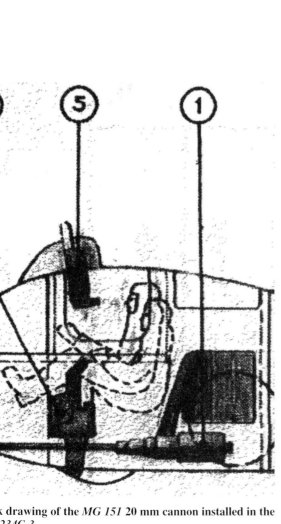

A pen and ink drawing of the *MG 151* 20 mm cannon installed in the nose of an *Ar 234C-3*.
#01 - one *MG 151* 20 mm cannon including its ammunition feed;
#06 - spend 20 mm shell case ejection chute.

An *MG 151* 20 mm cannon seen looking down its long barrel. A long barrel was needed due to the cannon's aft placement in the *Ar 234C*.

Arado Ar 234C

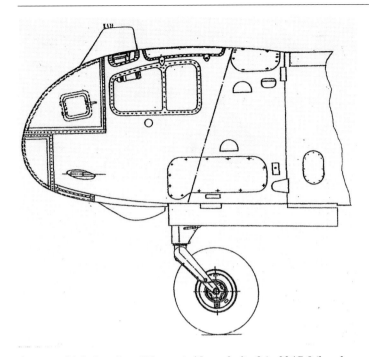

A pen and ink drawing of the port side cockpit of *Ar 234C-8* (bomber version) and featuring the location of its *MG 151 20* mm cannon. The cannon's barrel (in the cannon shroud) extends beyond the fuselage but not by much. The cannon shroud is seen as the flatten oval in the drawing. The access hatch to the cannon's operating mechanism and ammunition feed belt is behind the rectangular-shaped hatch. Pen and ink drawing by *Günter Sengfelder*.

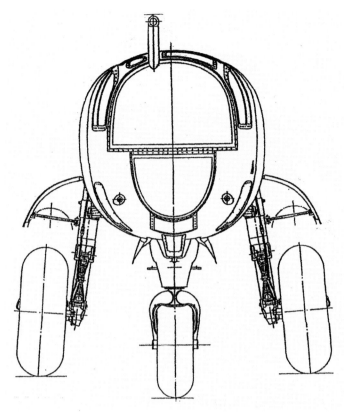

A pen and ink illustration of a nose-on view of the *Ar 234C-8* bomber version with a pair of 2x*MG 151 20* mm cannon in its nose. Pen and ink drawing by *Günter Sengfelder*.

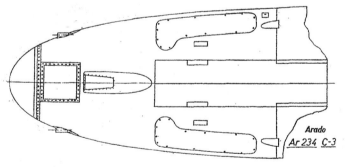

A pen and ink drawing of the underside of an *Ar 234C-3*'s fuselage and featuring the access hatches for each of its 2x*MG 151 20* mm cannon. Pen and ink drawing by *Günter Sengfelder*.

The access hatch to this *Ar 234C*'s MG 15*1 20* mm cannon has been removed. Its *MG 151* is still mounted inside, however, this poor quality photograph does not allow one to see it clearly. Note that the cannon's barrel does not protrude far beyond its shroud.

Arado Ar 234C

Ar234C-3

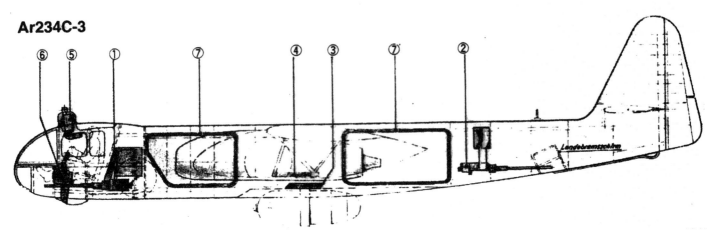

The *Ar 234C-3* were to have come equipped with a second pair of *MG 151 20* mm cannon aft in the tail section: port and starboard. This pen and ink drawings shows (#02) where the aft *MG 151's* would be placed.

The *MG 151 20* mm cannon's barrel does not extend much beyond its shroud as shown in this photograph of an *Ar 234C-3*. The *MG 151's* barrel can be seen directly below the horizontal stabilizer. Scale model and photographed by *Günter Sengfelder*.

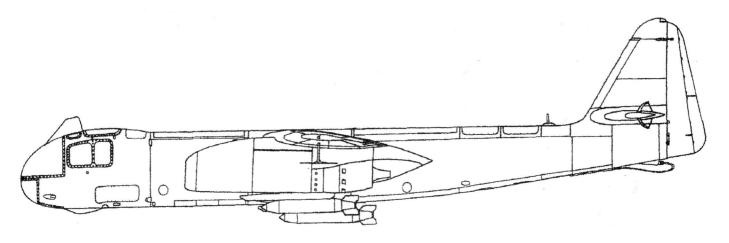

An *Ar 234C-3* with *3xR100 BS* unguided air-to-air rocket missiles

Ar 234C-3 Variations

Ar 234C-3s Carrying Unguided Air-To-Air Anti-Aircraft Rocket Missiles

Ar 234C-3W with 6xR1000BS Rocket Missiles

The air-to-air rocket missile (*R1000BS*) was a modification (*Rüstsatz*) added to the standard *Ar 234C-3* and known in the field as modification kit "*W*." This kit included 3x*R1000BS* unguided rocket missiles units containing 2x*R1000BS* each. The *R1000BS* rocket missile was approximately 5 feet 9 inches [1.8 meters] long and approximately 9 inches [210 mm] in diameter. Each missile had an incendiary warhead with 460 termite cylinders. Each rocket had a range of approximately 3/4 miles [1.2 kilometers]. They were mounted on the *Ar 234C-3* via *AG 140* racks...one under the center fuselage and under each dual engine nacelle. Total weight of each *RS1000BS* set of two rocket missiles was 1,488 pounds [675 kilograms].

Ar 234C-3 with 9x14 WK BS Rocket Missiles

This modification (*Rüstsatz*) included the 3x290 mm unguided air-to-air rocket missiles mounted on *AG 200* racks...one under the center fuselage and one under each dual engine nacelle. Total weight of the three of *14 WK BS* rocket missile packs as fitted to a *Ar 234C-3* was 1,800 [815 kilograms] pounds.

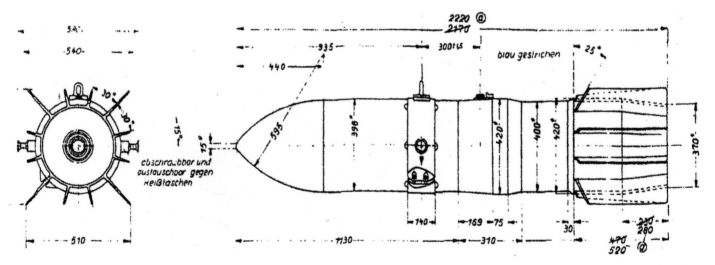

External appearance and overall size of an *R100 BS* unguided air-to-air rocket missile of the type to be carried by *Ar 234C3s*. Each *R100 BS* was approximately 5 feet 9 inches long with a range of 3/4 miles.

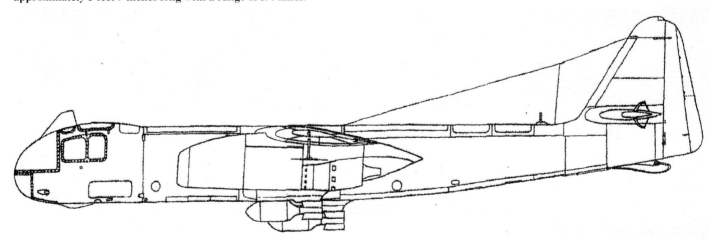

An *Ar 234C-3* with *9x14WK BS* unguided air-to-air rocket missiles.

Arado Ar 234C

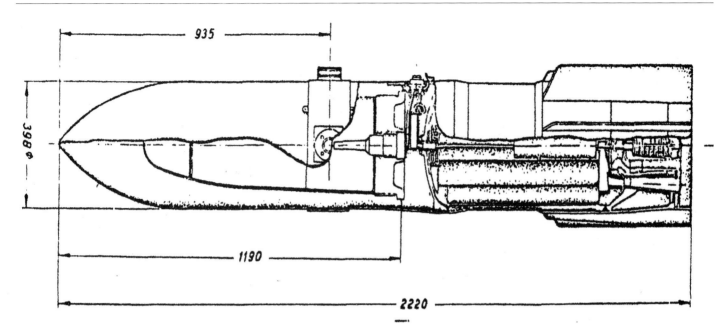

Internal arrangement of the *R100 BS* unguided air-to-air rocket missile.

Ar 234C-3 with WGr.21 Rocket Missiles

Ar 234C-3s were to be modified to carry three containers holding 4x*WGr. 21* centimeter rocket missiles. The *WGr. 21* was a unguided air-to-air anti-aircraft rocket missile which was fired from a distance into a Allied bomber pack. One container was located under the fuselage. In addition, two other containers with hung from racks under each of the two turbojet engine nacelles. Total weight of this modification (*Rüstsatz*) was approximately 4,000 [1,800 kilograms] pounds. At this weight, the *Ar 234C-3* was overloaded and would have required 2x*HWK 500* or *501* bi-fuel liquid rocket take-off assistance units.

This is how it was to have been: two *Me 110's* attack an Allied bomber formation with *WK BS* rocket missiles.

It took up to three armor personnel, often four, to load a single *WGr. 21* centimeter rocket shell into its tube on this *Me 410 B-1*. This was because each rocket weighed 242 pounds and was 49 1/4 inches long.

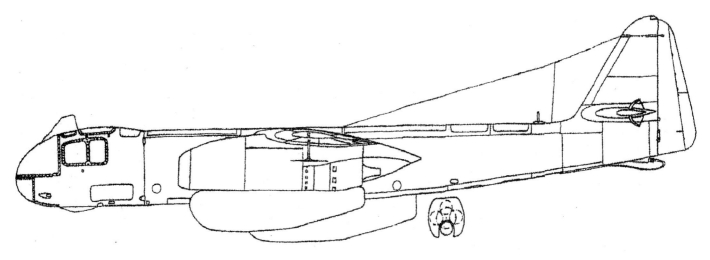

Ar 234C-3 with *210 mm WGr. air*-to-air rocket missiles.

The *Rüztsätze* modification kit for the *Ar 234C-3* would have included up to three containers holding four 21 centimeter *WGr.* rocket shells in a revolving chamber. This pen and ink drawing is how the *Rüztsätze* container with its revolving chamber would have looked on the *Ar 234C-3*.

This *Me 410 B-1* was modified to carry a revolving drum containing 6x*WGr.* 21 centimeter rocket shells. This photograph was taken about 1944. Later it was discovered that the fuselage suffered considerable damage from this arrangement and the scheme was abandoned. A single *WGr. 21* rocket shell had a maximum range of 25,754 feet [7,850 meters] or nearly five miles.

Arado Ar 234C

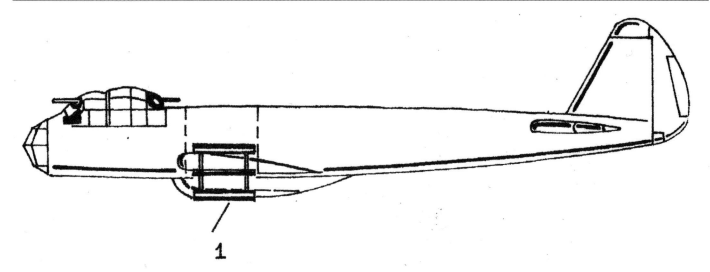

A proposed twin engined *Luftwaffe* night fighter with a *Rüztsätze* 21 centimeter *WGr.* modification kit.

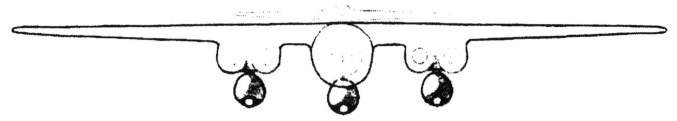

Bewaffnung: 4-12×21cm Wgr. 42 in 1-3 Vierlingswerfern

A pen and ink drawing of how the *Rüztsätze*-modified *Ar 234C-3* with container pods holding 21 centimeter *WGr.* rocket missiles would have appeared from out in the front of the turbojet-powered flying machine.

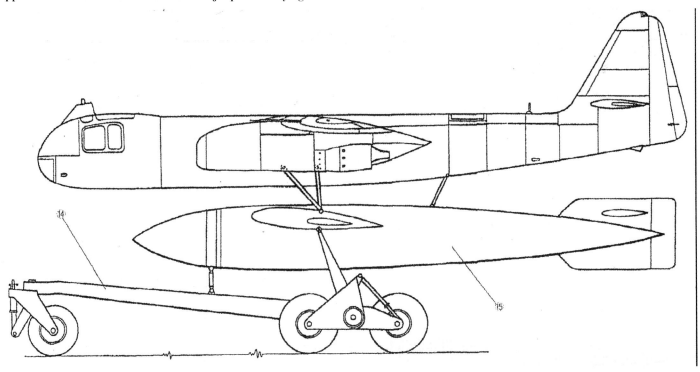

Pen and ink drawing by *Arado* of their *Ar 234C-3* with the proposed *Ar E.377* glide bomb and the whole combination known as the *Mistel 6.**

Ar 234C-3 with Ar E.377

The *Ar 234C-3* was the carrier aircraft for the unmanned *Arado Ar E.377* glide bomb. The *E.377*, which had been designed to fit beneath a *Ar 234C-3*, was a small, cylinder-shaped, *HWK* bi-fuel liquid rocket engine-powered, with high aspect ratio wings. This machine, which *Arado* was calling the "*Mistel 6*," had been developed from the *Luftwaffe's* 3,968 pound [1,800 kilogram] bomb. The *Ar 234C-3* and its *Ar E.377* would have been placed on a take-off trolley. A trolley of the type to be used with the *234C-3/E.377* combination had already been tested by *Heinkel, AG* for their *He 162* "*Mistel 5*" project. This was the 1,800 kilogram bomb powered by two turbojet engines. None the less, the *Ar E.377* was still just a project design by *Arado* engineers and no prototype was constructed. The *Ar 234C-3* carrier aircraft would have been overloaded with the *Ar E.377* attached and therefore would have required as many as 4x*HWK 500* or *501* bi-fuel liquid rocket take-off units.

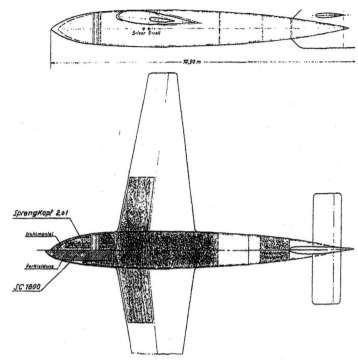

A pen and ink drawing by *Arado* of their proposed 3,968 pound *Ar E.377* unmanned glide bomb.

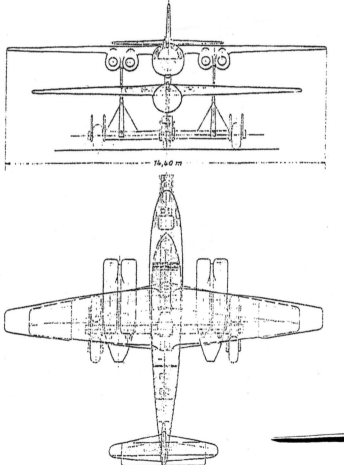

Below: A pen and ink illustration of *Arado's* proposed *Ar E.377* unmanned glide bomb as seen from its port side after being released by an *Ar 234C-3*.

A pen and ink drawing by *Arado* of the *Ar E.377* unmanned glide bomb perched atop an *Ar 234C-3* as seen from the front and beneath.

Arado Ar 234C

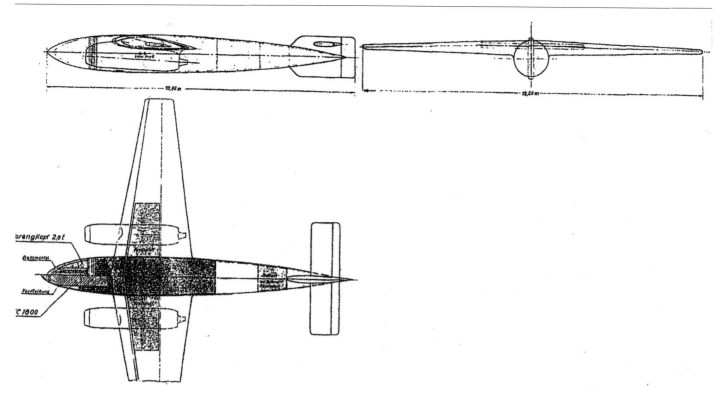

A pen and ink 3-view drawing by *Arado* of version #2 of their unmanned *Ar E.377* glide bomb. This version was to have been propelled by 2x*BMW 003A-1* turbojet engines. As a powered bomb the *Ar 234C-3*'s bombardier would have more control, therefore better accuracy in hitting the target.*

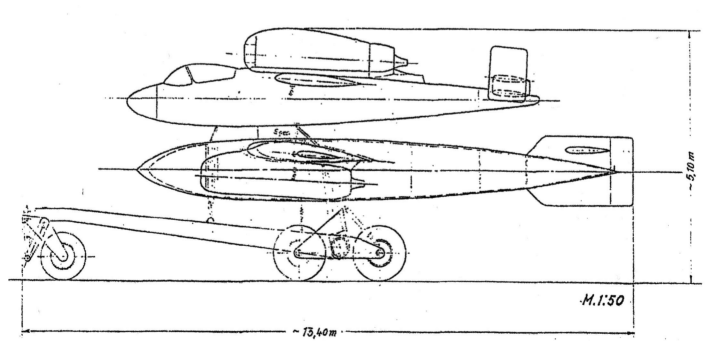

A pen and ink drawing from *Arado* featuring a *Heinkel He 162* atop an *Ar E.377*, unmanned but powered by 2x*BMW 003A* turbojet engines. Both are fitted to a three-wheeled take off dolly. *Arado* engineers were considering a similar arrangement for their *Ar 234C-3/Ar E.377*.*

Arado Ar 234C

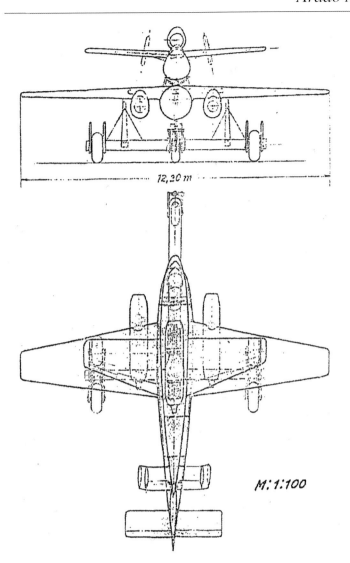

Ar 234C-3 with Ar E.381 Fighter/Interceptor

Arado had plans to modify one of their *Ar 234C-3s* as a carrier aircraft for the prone-piloted *Arado Ar E.381*. The *E.381*, which had been designed to fit beneath a *Ar 234C-3*, was a small, cylinder-shaped, *HWK* bi-fuel liquid rocket engine-powered high altitude fighter/interceptor. This machine was only a project design by *Arado* engineers and no prototype was constructed. Its wing span was approximately 29 feet and its overall length was about 18½ feet. Armament of the *E.381* included 1x*MK 108* 30 mm cannon located beneath the fuselage's upper surface at its center section. The *E.381*'s small size did not allow for a lot fuel to be carried. So, like the *Me 163*, the *E.381* probably could only make two passes at American *B-17* bomber formations before having to return to its base. It would land on a built-in skid similar but much smaller than that found on the *Me 163*. It's likely that the *E.381* would have faced the same challenges the *Me 163* experienced when approaching its landing area. U.S.A.A.F. fighter planes waited overhead as the defenseless and powerless tiny machines glided in for landing...shooting them out of the sky.

A pen and ink drawing from *Arado* featuring an *He 162* atop a 2x*BMW 003A* powered unmanned *Ar E.377* as seen from the front and above. *Arado* engineers wanted to substitute the *He 162* for their *Ar 234C-3*.*

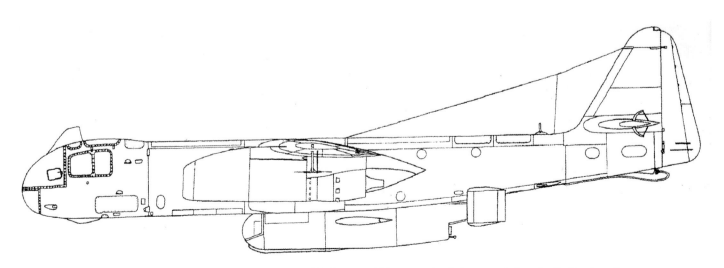

A pen and ink drawing by *Arado* featuring their *Ar 234C-3* with the *Ar E.381*, a parasite fighter carried under its fuselage.*

Arado Ar 234C

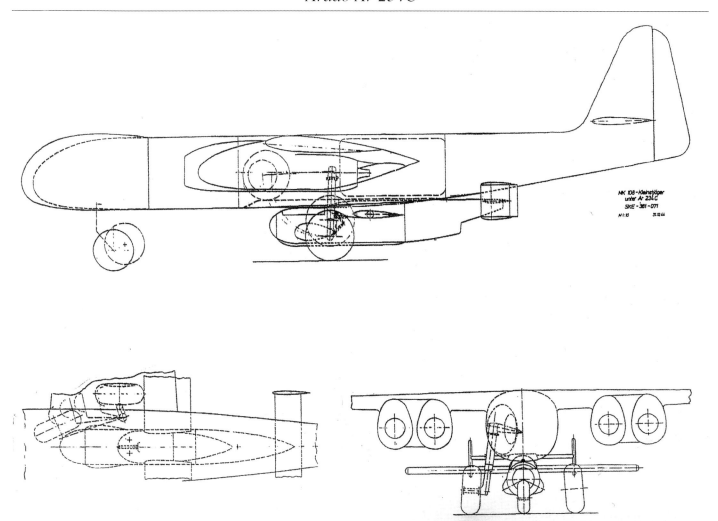

Pen and ink drawings of the *Ar 234C* carrying the single seat *Ar E.381*, *HWK 509C-1* bi-fuel rocket powered miniature fighter/interceptor. Top illustration is a ground level port side view with the *Ar E.381* attached. The illustration on the right is a nose on view of the *Ar 234C* with the *Ar E.381* carried beneath is fuselage. The illustration on the left is a view looking down at the *Ar E.381* from inside the *Ar 234C's* fuselage. Pen and ink drawing by *Günter Sengfelder*.

A nose-on port side view of the proposed *Arado* miniature fighter known as the *Ar E.381.1*. Landing the tiny aircraft would be achieved with the help of a wooden skid beneath the fuselage. Scale model and photographed by *Günter Sengfelder*.

Arado Ar 234C

A view of the underside of an *Ar 234C-3's* tricycle landing gear arrangement as it passes overhead. Notice the narrowness of the landing gear on the all of the *Ar 234* versions. Scale model and photographed by *Günter Sengfelder*.

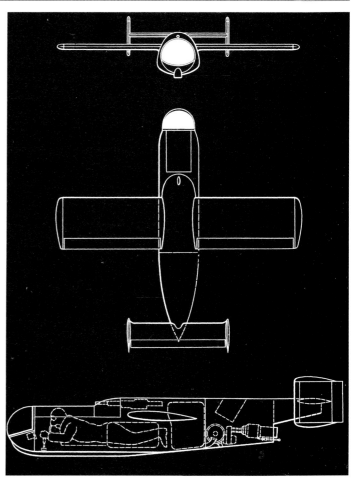

A three-view pen and ink drawing of the proposed *Ar E.381 HWK 509C-1* bi-fuel liquid rocket fighter/interceptor. The *HWK 509C* was to have provided 3,740 pounds [1,700 kilograms] thrust. Unlike the *Bachem* "Natter" or *Heinkel's* "Julia," which were to be ground launched, the *Ar E.381* would have been carried close to its operational altitude beneath an *Ar 234C*.

A port side view of the Ar E.381. Scale model and photographed by *Günter Sengfelder*.
• Engine - 1x*HWK 509C's* lower "cruising" chamber providing 660 pounds (299 kilograms) thrust.
• Wing span - 16 feet.
• Length - 18 feet 7 inches.
• Wing area - 59 square feet.
• Weight, empty - 1,963 pounds.
• Weight, takeoff - 2,690 pounds.
• Speed, maximum - 559 mph (900 km/h)

Arado Ar 234C

The port side of an *Ar 234C* loaded with a single *Ar E.381* under its fuselage between the main landing gear. Notice the very little space between the ground and the *E.381*. The miniature fighter was occupied by its pilot at the time of the *234C's* takeoff and it is doubtful that the miniature fighter would survive were the *234C* to abort shortly after lift off. Scale models and photographed by *Günter Sengfelder*.

A nose-on view of the *Ar 234C* with its *Ar E.381* appearing behind the nose wheel. Scale models and photographed by *Günter Sengfelder*.

A port side rear view of the *Ar 234C* loaded with a single *Ar E.381* miniature fighter. Notice the widely spaced dual rudders on the miniature fighter. This glide bomb was similar in length and diameter to the *Fi 103 (FZG 76)* flying bomb, however, its single pulse-jet engine had been removed because it wasn't needed on a glide bomb. Scale models and photographed by *Günter Sengfelder*.

An overhead tail starboard side view of the *Ar 234C* with the *Ar E.381* attached, although it is barely visible. Scale models and photographed by *Günter Sengfelder*.

A ground level view of the *Ar 234C* and its *Ar E.381* miniature fighter on a dolly parked out front. Scale models and photographed by *Günter Sengfelder*.

Nose port side view of the *Ar 234C* with the *Ar E.381* seen attached beneath. Scale models and photographed by *Günter Sengfelder*.

Arado Ar 234C

A close up detail of the *Ar 234C-3's* gear. This machine is carrying an *Arado* miniature *HWK* bi-fuel liquid rocket propeller fighter...the *Ar E.581*. Scale models and photographed by *Günter Sengfelder*.

A pen and ink drawing from WW2 showing a proposed *Ar E.381* miniature fighter shortly after its release from an *Ar 234C-3*.

A close-up view of the *MK 108* 30 mm cannon as seen from its port side.

The *Maschinenkanone (MK) 108 30* mm cannon by *Rheinmetall-Borsig*. The installation of *MK 108* cannon was planned for several *Ar 234C* variations. Notice the size of the individual *MK 108* 30 mm cannon shell.

Arado Ar 234C

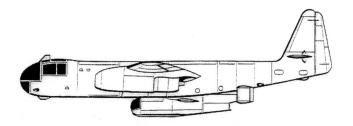

Ar 234 C-3 - Ar E 381/I

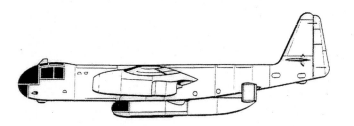

Ar 234 C-3 - Ar E 381/II

Ar 234C-3 - Fi 103V1 Missile Carrier "Huckepack"

About October 1944, *Arado* began making plans to modify *Ar 234C-3s* to carry a single *Fi 103 V1 (FZG 76)*. One approach, using the *Ar 234C-3*, was to place the *Fi 103* on top of the *C-3*'s fuselage and secure it with hydraulic arms fixed to the *C-3*. Prior to launching the *Fi 103*, the hydraulic arms would be raised by the pilot so that the pilotless machine would clear the *Ar 234C-3*'s tail assembly.

A second approach under consideration was to hang the *Fi 103V1* beneath the *Ar 234C-3* and place it all on a three-wheeled take-off trolley (*startwagen*) similar to the trolley used when the *Ar 234 V6* and *V8* were being flight tested with a landing skids. *Rheinmetall-Borsig* was designing the three-wheeled take-off trolley for use in "*Huckepack*" operations. It is unknown whether testing was carried out...perhaps not.

Another *Arado 234C* variation included an *Ar 234C-3* carrying under its fuselage, a miniature bi-fuel liquid rocket powered fighter up to altitudes where the Allied bomber stream preferred to fly. The most common miniature fighter was the *Arado*-designed *Ar E.381.1*. This was *Arado*'s offer to the *RLM* in August 1944, for a *HWK 509*-powered high-speed interceptor. *Junkers*' offered their *EF 127* "Dolly," Heinkel their *He P.1077*, and *Bachem* his "*Natter*." The "*Natter*" won the day given heavy lobbying by *Reichsführer SS Heinrich Himmler*. The *Ar E.381* differed from the "*Natter*," the "*Julia*," and Junkers' *EF 127*, because it was the only machine proposed to be carried aloft by another aircraft. A second version of the *Arado* miniature fighter was known as the *Ar E.381.2*. Very little is known about this miniature fighter version.

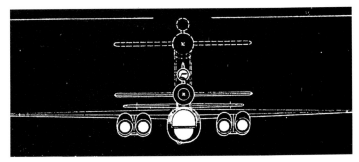

An alternative arrangement for taking an *Fi 103* up to altitude and/or sufficient forward motion to get its pulse-jet working was to tow it behind the *Ar 234C*. This towing method was known as "*Deichselschlepp*."

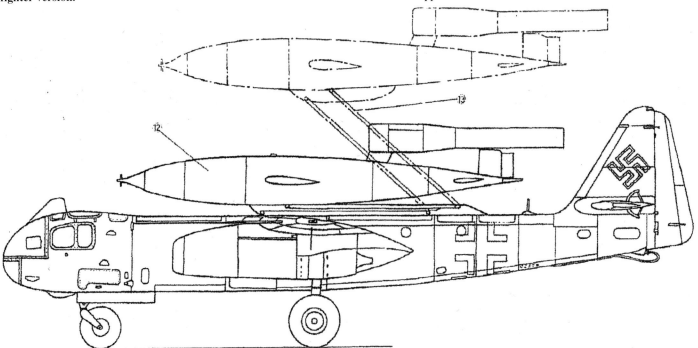

The *Ar 234C-3* carrying an *Fi 103 V1 (FZG 76)* guided missile above its fuselage.

Arado Ar 234C

A combination involving the *Ar 234C-3* and the *Fi 103 V1* designed to bomb England. It would have most likely succeeded. The *V1* was 25 feet 4 inches long, constructed of plywood and sheet steel. The direction of the *V1* was generated automatically by an on board gyroscope which gave signals to the ground. Returning signals, in turn, operated the rudder and elevator. Directional information was obtained from a pre-set compass. An air log measured the pre-set distance after which the elevators were depressed and the *V1* descended to the ground. The nose of the *V1* held 1,874 pounds [850 kilograms] of high explosive, bolted to the front of the fuel tank. Wing span was 17 1/2 feet, and it was propelled by a single *Argus* pulse-jet engine.

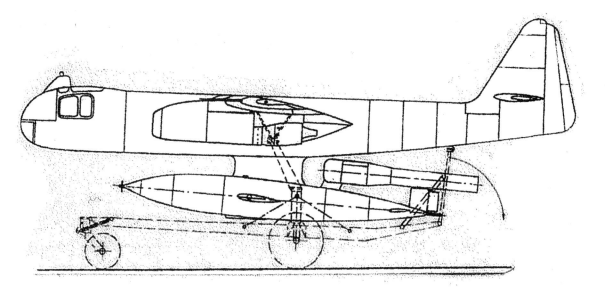

Arado was considering an alternative method of carrying the *Fi 103*. In this proposal the *Fi 103* would be hung from beneath the fuselage of the *Ar 234C-3* and then placed on a three wheeled dolly for take off. This arrangement was similar to the way *Arado* got the *Ar 234(C)V6* and *V8* prototypes off the ground.

Ar 234C-3 - Fi 103V1 Missile Tower "Deichselschlepp"

About October 1944, *Arado* was making plans to modify *Ar 234C-3s* to tow a single *Fi 103 V1 (FZG 76)*. Several approaches were considered. In one case, *Arado* engineers had taken the *Fi 103V1* and placed it on its own fixed two-wheeled trailer and connected to the *Ar 234C-3* by a long boom. The would be a pivot point on the underside of the *C-3*'s rear fuselage where the boom was secured. This method of towing had been tested by *DFS* engineers. They had built a engineless *Fi 103* known as the *SG 5041 (Sonderergerät)*. It's wheels were enclosed in spat/covers similar to the *Ju 87* "Stuka." Modifications were made to the *Ar 234C-3*, too.

The main change involved in adding a small plexiglass-covered compartment in the *C-3* where the test engineer sat rearward observing the experimental towing operation. This compartment was aft the wing's trailing edge. Tests began on 9 February 1945. The *Ar 234C-3* took off and landed with the *SG 5041V1*. The first test was a success. During one later test, the *SG 5041V2* tore itself from the *Ar 234C-3* due to excessive vibrations and the *SG 5041V2* crashed. It appears that testing continued up until war's end although excessive vibrations of the towed item remained pretty unsolved.

Arado Ar 234C

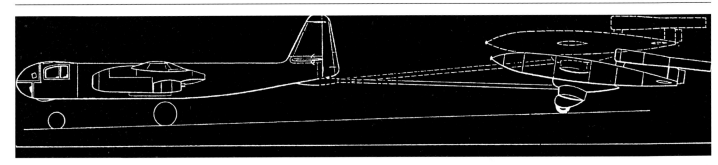

A pen and ink drawing by *Arado* featuring their *Ar 234C-3* towing an *Fi 103 V1* missile...*Deichselschlepp* style. In this version, the *Fi 103* still had its *Argus* pulse-jet power plant and the only difference between this and other methods was that the *Fi 103* was towed...not carried.*

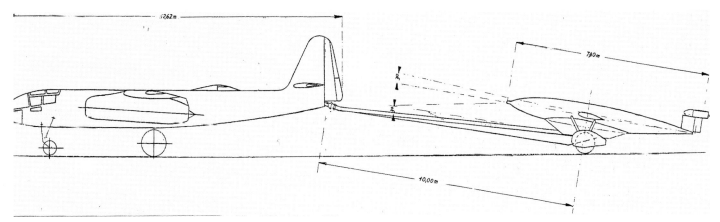

A pen and ink drawing from *Arado* featuring their *Ar 234C-3* towing an unpowered *Fi 103* missile which turned it into a gliding bomb.*

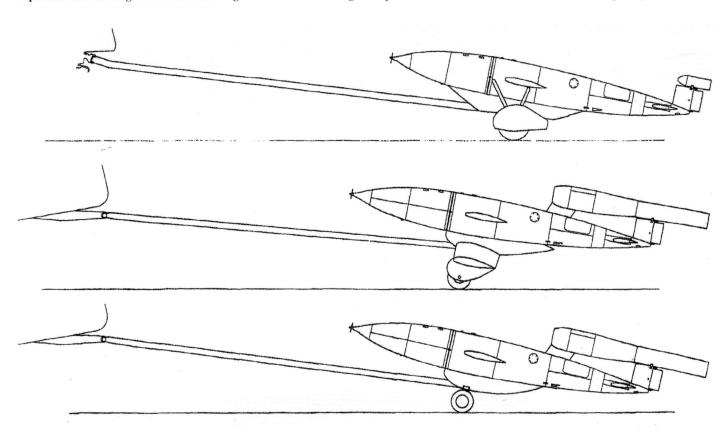

A pen and ink drawing from *Arado* featuring the several ways of towing *Fi 103*'s off the ground...both powered and unpowered. The *Fi 103* at the top and center have fixed towing gear. The wheels on *Fi 103* at the bottom would drop off once the *Ar 234C-3* and the *Fi 103* were airborne.

Arado Ar 234C

A poor quality photo of an unpowered *Fi 103 (FZG 76)* gliding bomb from its port side showing its fixed and spatted take off wheels.

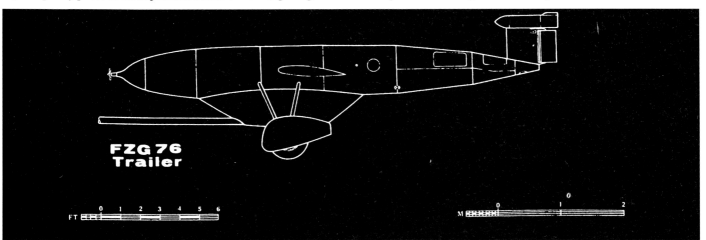

Arado suggested, too, that the *Fieseler Fi 103 (FZG 76)* bomb, minus its pulse-jet engine, be towed by an *Ar 234C* on a long boom attached to a two-wheel trailer. Upon lift off and airborne the trailer would fall away.

A poor quality photo showing an *Ar 234B* towing the *FZG 76* trailer during experiments. The arrangement would have been the same for an *Ar 234C*.

Ar 234C-3 with Towed Winged-Fuel Tank

Success gained with using the *Ar 234C-3* as a towing machine, resulted in experimenting with other towed items such as a external fuel tank. *Arado* engineers fitted a standard external gasoline tank of 66 gallons with 16½ [5.05 meters] foot wings. Others were tested with slightly larger wing spans of 20½ [6.25 meters] feet. The towing boom contained fuel lines through which fuel could be pumped out of the container during long-distance aerial photograph reconnaissance or bombing runs. With the success of the towed fuel tank, *Arado* engineers were considering plans to tow various oversized bombs and missiles.

A pen and ink illustration by *Arado* featuring one of their *Ar 234C-3s* in flight towing a winged fuel tank.

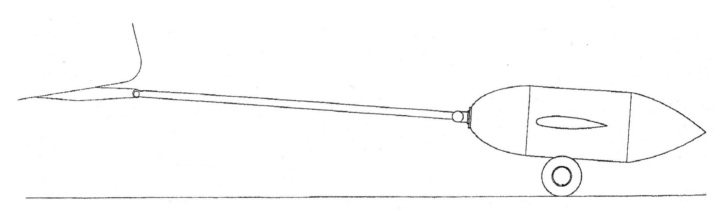

A pen and ink drawing by *Arado* showing how a towed, winged, fuel tank would be attached to an *Ar 234C-3's* rear fuselage.*

A *Deichselschlepp* towed bomb (*SC 500* or *SC 1000*) which has been fitted with short, stubby wings borrowed from an *Fi 103*, behind the *Me 262 V10* coded *VI+AE*.

The flexible connection/coupling between the towed *Deichselschlepp* and the tail of the *Me 262 V10*. A similar 13 foot [4 meter] towing bar arrangement was being considered using the *Ar 234C* as the towing vehicle.

Arado Ar 234C

The *Me 262 V10* has lifted off with its towed *SC 500* bomb.

An overhead photograph of the *Me 262 V10*. Everything appearing to be going nicely and was considered a success.

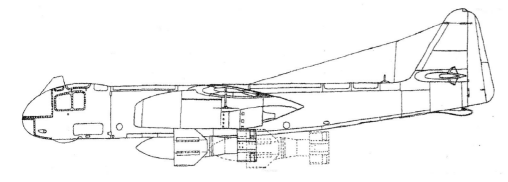

A pen and ink drawing from *Arado* featuring their *Ar 234C-3* carrying 2x *SD 1400 "Fritz-X"* air-to-ship guided armor-piercing bombs.

Ar 234C-3 with *"Fritz-X"* Air-To-Ship Guided Armor-Piercing Bomb

The *Ar 234C-3* was to be the next launching platform for the *SD 1400-X* (*Fritz-X*) guided high-explosive bomb for use against ground and naval targets. Its guidance and control came about through impulses transmitted through stranded wire from the parent aircraft, in this case the *Ar 234C-3*. The bomb weighed about 3,086 pounds [1,400 kilograms] and carried an explosive charge of 772 pounds [350 kilograms]. It was about 10½ feet [3.2 meters] long and had two short wings, each with a wing span of 10½ feet [3.2 meters]. The *"Fritz-X"* had become operational in Spring 1943 and had been tested on aerial launching platforms such as the *He 177*, however this aircraft was too slow and undependable to build a fleet carrying the *"Fritz-X."* Hopes were high that the *Ar 234C-3* would be the ideal launching platform for the *"Fritz-X."* The plan for attacking Allied ships would have required the *"Fritz-X"* carrying *Ar 234C-3* to fly over the target at considerable height so it could be out of reach of light and medium anti-aircraft fire. Bombing altitude was between 16,400 to 23,000 feet [5,000 to 7,000 meters]. Tests showed that releases of the *"Fritz-X"* from 16,400 feet achieved about 60% accuracy. The nearly new Italian battleship *"Roma"* was sunk by three direct hits from *"Fritz-X"* bombs on 9 September 1943, off the coast of Sardinia. The *Roma's* escort, the Italian battleship *"Italia"* was hit, too, by one *"Fritz-X."* Heavily damaged it did not sink.

The starboard side of a *"Fritz-X"* guided missile.

The *PC 1400 "Fritz X"* was a highly effective rocket powered anti-shipping glide bomb. Once its release from its parent aircraft, an *Ar 234C*, for example, the *PC 1400* could be guided down to its target...a war ship or a troop transport...where upon striking it, it would explode upon contact.

Arado Ar 234C

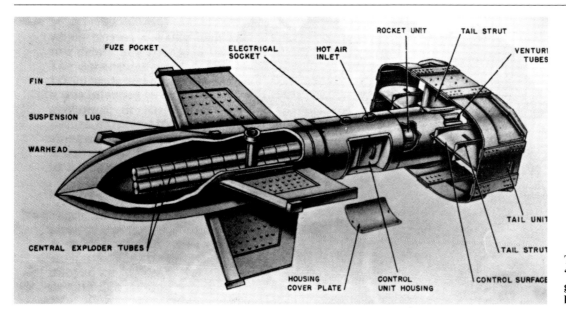

The insides of the *PC 1400* "*Fritz X*" rocket powered guided anti-shipping missile to be carried by *Arado Ar 234Cs*.

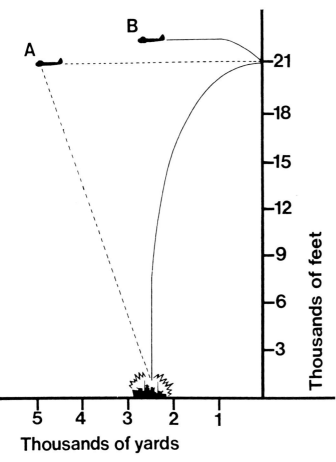

ATTACK WITH FRITZ X GUIDED BOMB

This pen and ink drawing shows just how an attack on a surface ship would be carried out with a *PC 1400* "*Fritz X*" bomb. At about 21,000 feet altitude the parent aircraft, such as an *Ar 234C*, would release the guided anti-shipping bomb. In its downward guiding angle, the *PC 1400* would have traveled about 7,500 feet horizontally from its point of release to its point of attack.

This fine artwork by Alfred Johnson illustrates how the *Luftwaffe* parent aircraft (*Do 217*) flying high over the "*Roma*," released its *PC 1400*, and guided it to the enemy target on the water's surface with the help of its rocket engine.

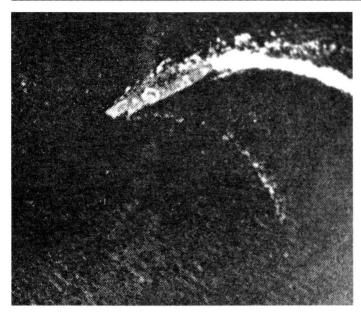

A close-up view of the Italian battleship "*Roma*" as it bares hard to port in a useless attempt to evade the approaching "*Fritz-X*."

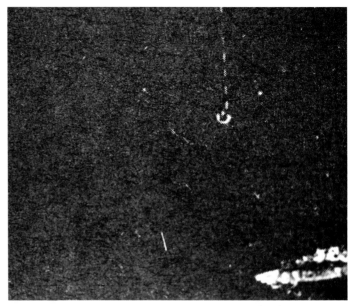

The white dotted line is the flight path of the incoming "*Fritz-X*." The half moon is the missile, and in the lower right-hand corner is the "*Roma*."

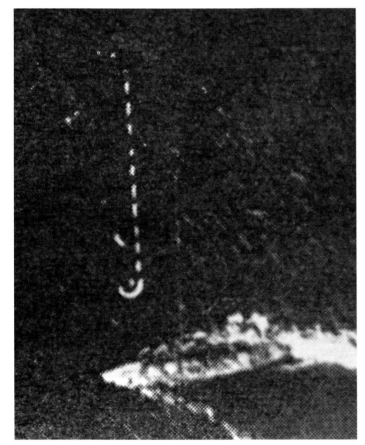

The "*Fritz-X*" is moments away from striking the "*Roma*" near its bow.

The death of the Italian battleship "*Roma*" hit by three "*Fritz-X*" guided missiles somewhere between Corsica and Sardinia on 9 September 1943.

Arado Ar 234C

Ar 234C-3 with "Panzerblitz" II PB Rockets

The *Panzerblitz PB* "*Rüstsatz*" modification was a *Ar 234C-3* equipped with three external container/pods holding 20x88 mm *PB II* air-to-ground rockets. One container was under the fuselage while others were under each of the dual turbojet engine nacelles. The *PB II* was a simple rocket with a armor-piercing head designed to be fired from a *Luftwaffe* fighter. The warhead was the same as 8.8 centimeter hollow-charge head as used on the *Panzerschreck* infantry rocket launcher. It was fitted with the rocket motor from the *R4M* and reached a velocity of 1,214 feet per second [370 meters]. It had been developed by *Waffenwerke Brünn* in Czechoslovakia, primarily for the *Henschel Hs 132*, which never was ready by war's end. Total weight of one *PB II* container, its rack, plus its full load of 20x88 mm rockets was 342 pounds [155 kilograms].

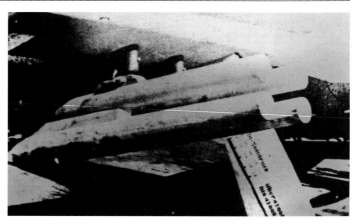

A group of *4xPanzerblitz-2* rocket shells hung under the starboard wing of a *Fw 190*.

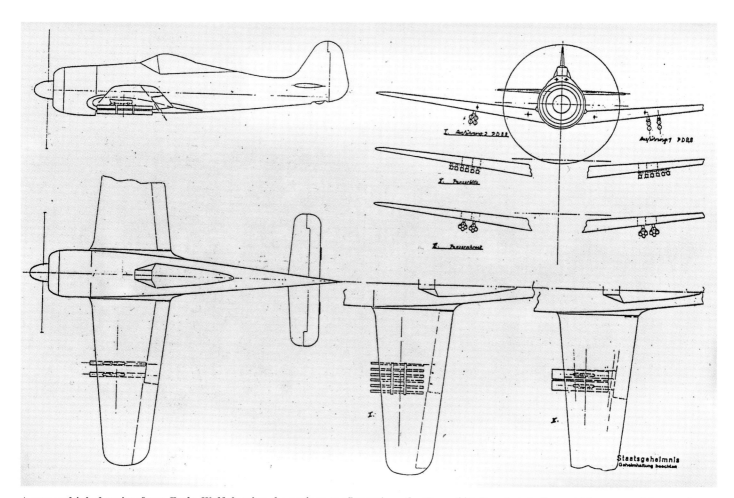

A pen and ink drawing from *Focke-Wulf* showing the various configurations the *Panzerblitz-2* rocket shells could be grouped for maximum carrying capacity on a *Fw 190*. For the *Ar 234C-3*, *Arado* had a revolving chamber containing up to 20 *Panzerblitz-2* rocket shells. The *Ar 234C-3* would be able to carry three of these containers...one under the fuselage and one under each turbojet engine nacelle.

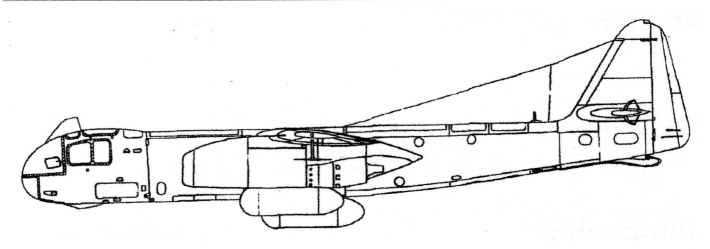

A pen and ink drawing by *Arado* featuring their *Ar 234C-3* with three individual containers of "*Panzerblitz*-2 *(Pb2)* air-to-ground rockets with high explosive war heads.

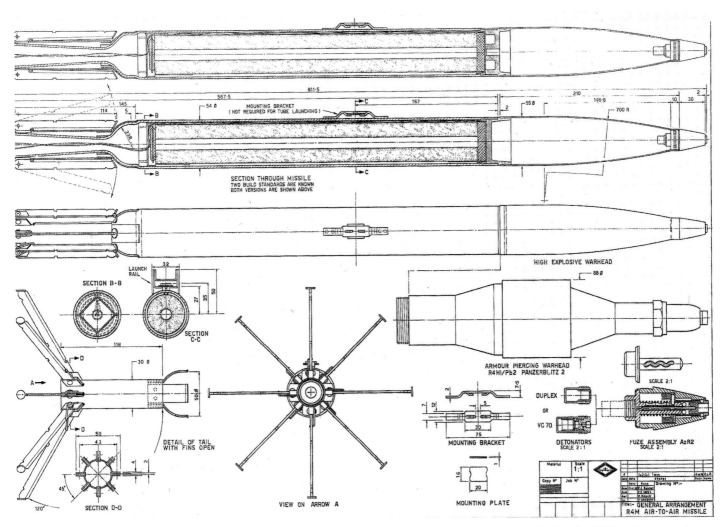

General arrangement drawing of the *R4M* air-to-air missile, and when the high explosive head was attached to the *R4M* rocket motor it was known as the *Panzerblitz-2 (Pb2)*.*

Arado Ar 234C

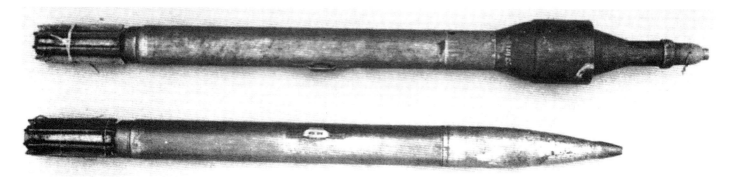

Top: the *Panzerblitz-2 (Pb2)*. Bottom: the *R4M* rocket shell.

A rack of *R4M* air-to-air rocket shells shown under the starboard wing of an *Me 262A*.

Ar 234C-3 "Heeresflugzeug"
Army Ground Support Aircraft

One version of the *Ar 234C-3* was to be a heavily armored ground attack/support machine called the "*Heeresflugzeug*." This single-seat *C-3* version would have carried a variety of air-to-ground rocket missiles. This included three 4xbarrel *Flak 42* missile launchers, *SC 500RS* rocket bombs, and 3x*SD4* anti-personnel bomb containers with 74 bombs in each container. In addition, the "*Heeresflugzeug*" was equipped with two nose-mounted *MG* 151 20 mm cannon and bomb racks to carry up to 3,310 pounds [1,500 kilograms] of conventional bombs. No *Ar 234C-3* version of the "*Heeresflugzeug*" had been built by war's end.

Below: A pen and ink drawing from *Arado* featuring their *Ar 234C-3* "*Heeresflugzeug*," or army ground support aircraft, when fitted out with *Flak 42* missile launch containers.*

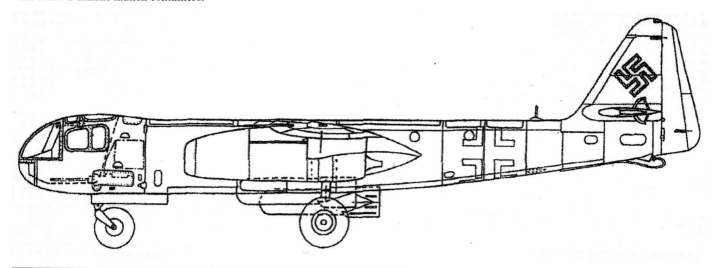

Arado Ar 234C

This is a photo of the nose of the wooden cockpit mockup for the proposed *Ar 234C-3 "Heeresflugzeug"* machine.

The wooden cockpit mockup for the proposed *Ar 234C-3 "Heeresflugzeug"* as seen from behind.

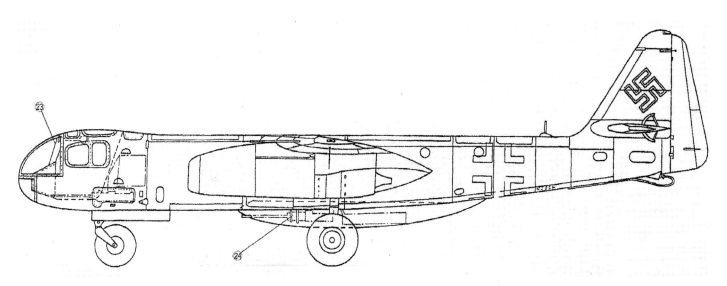

A pen and ink drawing by *Arado* featuring their proposed *Ar 234C "Hohenjäger"* or high altitude fighter/interceptor.*

Arado Ar 234C

Ar 234C-3 "Höhenjäger" High Altitude Fighter/Interceptor

One version of the *Ar 234C-3*, suggested by the *RLM* in May 1944, was a machine called the "*Höhenjäger.*" This single-seat *C-3* machine, with a pressurized cockpit, would have intercepted and fought Allied bombers at altitudes up to 39,400 feet [12,000 meters]. A armored cockpit would have provided protection for the pilot from hostile cannon fire. It was anticipated that this machine would have carried a variety of air-to-air rocket missiles but its main armament would have been 4x*MG 151* 20 mm cannon: two fixed forward in a "*Rüstsatz*" *R1* under fuselage pod and two fuselage nose mounted fix forward...one on the port side and the other on the starboard side. No *C-3* version of the "*Höhenjäger*" had been built by war's end.

Ar 234C-3/Na - Radar-Equipped Night Fighter

This was a designated radar equipped night-fighter using the single-seat *Ar 234C-3* fuselage with its 4x*BMW 003A-1* turbojet engines. Two prototypes were to have been built...the *Ar 234C-3/Na* and the *Ar234C-3/Nb*. The *Ar 234C-3/Na* was to have had a small compartment built in the rear fuselage for the *FuG 218* airborne interception radar operator. External radar antenna included the old style, nose-mounted *Siemens Geweih* "stag antlers." Offensive armament included 1x*MG 151* 20 mm cannon fixed forward in the fuselage nose. In addition, the *Ar 234C-3/Na* featured the "*Rüstsatz*" *R3* under fuselage pod with 2x*MK 108* 30 mm fixed forward cannon and 100 rounds per cannon. The *R3* kit weighed 661pounds [300 kilograms]. Later, the designated night fighting version of the *Ar 234C* would not be the *234C-3/Na* or *Nb*, but the proposed *Ar 234C-7*. In this version, the pilot and the radar operator sat side-by-side in the cockpit. Cockpit nose glazing would be reduced...the upper half glazed while the lower half was covered over. However, a prototype of the *Ar 234C-3/Na* night fighter is believed to have been constructed at Sagan prior to war's end: the *Ar 234V23*. This is believed because *Arado* documents state that the *234V23* was to have been completed by 31 January 1945.

A pair of fixed fire *MG 151* 20 mm cannon set up in a *Rüstsätze* compartment/pod arrangement for use on the *Ar 234C-3*.

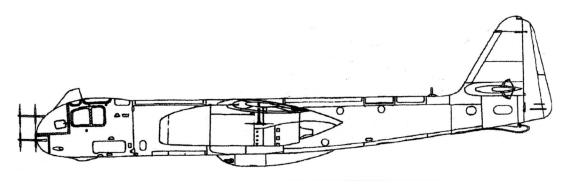

A pen and ink drawing by *Arado* of their proposed *Ar 234C-3/Na FuG 218* radar-equipped night fighter.*

Arado Ar 234C

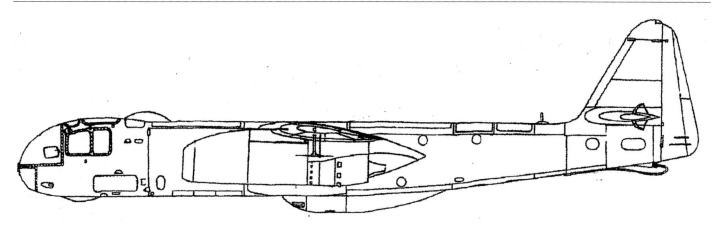

A pen and ink drawing by *Arado* of their proposed *Ar 234C-3/Nb FuG 244* radar-equipped night fighter.*

Ar 234C-3/Nb - Radar-Equipped Night Fighter

This was a designated radar equipped night-fighter using the single-seat *Ar 234C-3* fuselage with its 4x*BMW 003A-1* turbojet engines. Two prototypes were to have been built...the *Ar 234C-3/Na* and the *Ar234C-3/Nb*. The *Ar 234C-3/Nb*, like the Ar *234C-3/Na*, had a small compartment built in the rear fuselage for the *Siemens FuG 218 "Neptun V"* airborne interception radar operator. External radar antenna aerial array such as the old style, nose-mounted *Siemens Geweih* "stag antlers" used on the *Ar 234C-3/Na*, was eliminated. Instead, the *Ar 234C-3/Nb* had the *Telefunken FuG 244 "Bremen O"* centimetric wavelength interception radar. Its aerial was parabolic and the system operated on a 9 centimeter wavelength. In the *Ar 234C-3/Nb*, the radar's parabolic dish was housed beneath a clear plastic dome aft the cockpit canopy. Offensive armament included 1x*MG 151* 20 mm cannon fixed forward in the fuselage nose, the same as in the *Ar 234C-3/Na*. In addition, the *Ar 234C-3/Nb* featured the *Rüstsatz R3* under fuselage pod with 2x*MK 108* 30 mm fixed forward cannon and 100 rounds per cannon as found in the *Ar 234C-3Na*. The *R3* kit weighed 661 pounds [300 kilograms]. Later, the designated night fighting version of the *Ar 234C* would not be the *234C-3/Na* or *Nb*, but the proposed *Ar 234C-7*. In this version, the pilot and the radar operator sat side-by-side in the cockpit. Cockpit nose glazing would be reduced...the upper half glazed while the lower half was covered over. However, a prototype of the *Ar 234C-3/Nb* night fighter is believed to have been constructed at Sagan prior to war's end as *Ar 234V27 werk nummer 130066*. This is because *Arado* documents state that *werk nummer 130067*, was to have been completed by 28 February 1945. So, it was under construction, however, whether this night-fighting prototype was, in fact, completed by the time of the surrender, is not known for certain.

Ar 234C-3 with *FuG 240-4 "Bremen"* Rotating Radar Disk (AWACS)

This machine with the standard *C-3* fuselage, *C-3* wings, two-man crew, armament, and 4x*BMW 003A-1* turbojet engines was to carry the brand new *FuG 240-4 "Bremen"* centrimetric wave length interception radar. The *FuG 240-4* was unique in that inside its radome was a rotating radar unit looking similar to todays *AWACS* machines. The *FuG 240-4* was undergoing wind tunnel testing at war's end, featured a 180° observation angle which could detect and track enemy Allied bombers at night. Its rotating disk was enclosed in a 4 foot 9 inch [1.5 meter] flat dish-shaped radome and it would have been mounted on a single pylon above the wing's center section. *Arado* documents state that the *FuG 240-4* had a tracking range of between 4 and 5 miles [6,000 to 8,000 meters]. A 1:10 scale model of the complete *FuG 240-4* apparatus had been undergoing wind tunnel testing at *DVL*-Berlin/Adlershof since February 1945. No *Ar 234C-3* had been modified to accept the *FuG 240-4* unit even for flight testing.

Arado Ar 234C

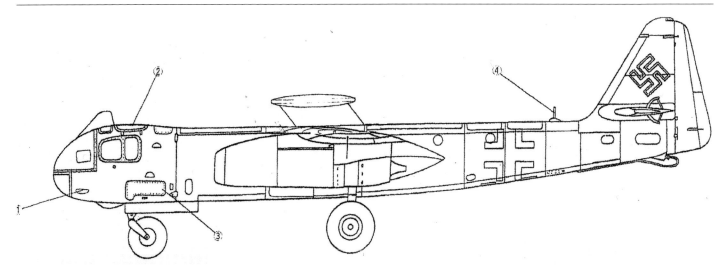

A pen and ink drawing by *Arado* of their proposed *Ar 234C-3* with the *FuG 240-4* "*Bremen*" rotating radar disk.*

Ground level port/nose side view of the *Ar 234C-3* with its *FuG 240-4* "*Bremen*" rotating radar disk.

A ground level direct port side view of the *FuG 240-4* rotating radar-equipped *Ar 234C-3* seen on a runway's edge near a wooden area. Scale model and photographed by *Günter Sengfelder*.

Arado Ar 234C

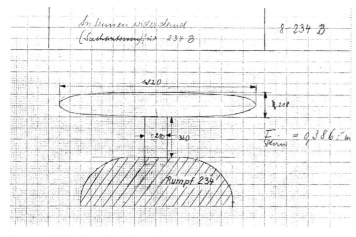

Pen and ink drawing featuring the dimensions of the *FuG 240-4* "*AWACS*" rotating radar disk. It measured about 8 1/2 inches (218 cm) thick and had a diameter of almost 6 feet (1,820 cm). The disk was located approximately 14 inches (360 mm) above the *Ar 234C-3*'s fuselage on a mount 8.3 inches (210 mm) thick. Pen and ink drawing by *Günter Sengfelder*.

A starboard/nose view of the *Ar 234C-3* with its *FuG 240-4* "*Bremen*" rotating radar disk. Scale model and photographed by *Günter Sengfelder*.

A port side/rear view of the *FuG 240-4* "*Bremen*" rotating radar-equipped *Ar 234C-3*. Scale model and photographed by *Günter Sengfelder*.

A view of the *FuG 240-4* "*Bremen*" radar-equipped *Ar 234C-3* as seen from its rear port side. Scale model and photographed by *Günter Sengfelder*.

Arado Ar 234C

An overhead view of the *Ar 234C-3*, *FuG 240-4 "Bremen"* radar-equipped four turbojet engine-powered machine. The *FuG 240-4* radar disk's shape had been wind-tunnel tested, however, none are known to have been installed and flight-tested on an *Ar 234C*. Scale model and photographed by *Günter Sengfelder*.

A close-up of the *Ar 234C-3* and its dorsal-mounted *FuG 240-4* rotating radar disk. Scale model and photographed by *Günter Sengfelder*.

A view of the *Ar 234C-3* with its *FuG 240-4 "Bremen"* as seen from its rear starboard side. Scale model and photographed by *Günter Sengfelder*.

Arado Ar 234C

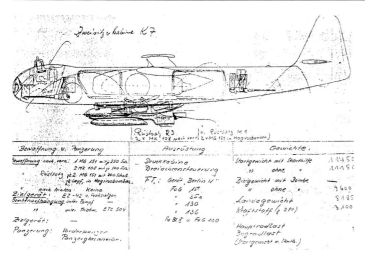

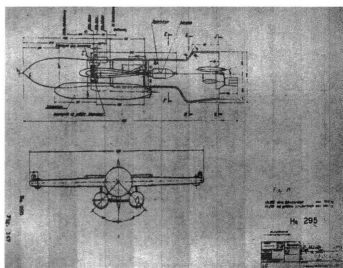

The proposed *Ar 234C-3*, *FuG 240-4* radar-equipped machine was also designed to carry up to three *Henschel Hs 295* anti-shipping guided missiles...one beneath its fuselage and two more under its wings. This crude, hand-drawn pen and ink drawing is from *Arado*. Date unknown. In addition to its *Hs 295* missiles, it would be heavily armed and defended: • 1x*MG 151* 20 mm fixed rear-firing cannon with 250 rounds; • 2x*MK 108* fixed forward-firing cannon with 100 rounds each; • "Berlin N" internal nose-mounted radar; • *FuG 15*; • *FuG 25a*; • *FuG 120*; • *FuG 130*; • *FuG 136*

A three-view pen and ink drawing of the *Henschel Hs 295* anti-shipping guided missile designed to be used on the *Ar 234C-3*. This missile was basically a *Hs 293*, however, *Henschel* had moved the twin rocket propulsion motors closer to the war head so that it would have more ground clearance when mounted under the fuselage of the *C-3*.

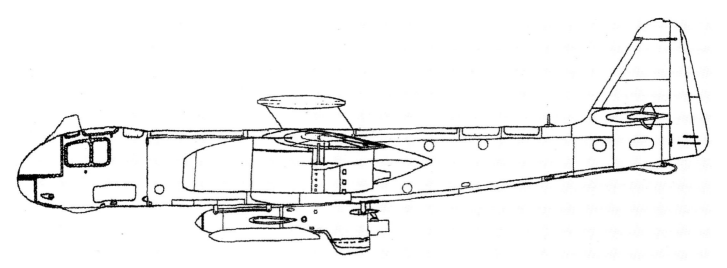

A pen and ink drawing from *Arado* of their proposed *Ar 234C-3*, *FuG 240-4* rotating radar, and *Hs 295* anti-shipping guided missile carrier.*

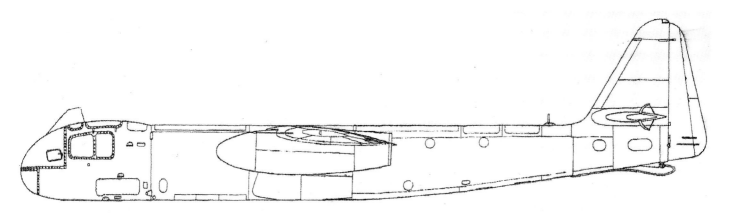

A pen and ink drawing by *Arado* of the port side of their *HWK 509A* bi-fuel liquid rocket propelled *Ar 234C-3R* photo reconnaissance aircraft.*

Arado Ar 234C

Ar 234C-3R...HWK Bi-fuel Liquid Rocket Powered Reconnaissance Aircraft

About March 1944, *Dipl.-Ing. Wilhem van Nes*, of *Arado's* design group, proposed a *234C-3* in response to the *RLM's* call for a high altitude reconnaissance aircraft. The *RLM* was seeking a machine capable of operating at 75,458 [23,000 meters] feet or higher. *Van Nes* had several ideas. One version was to replace the *234C-3*'s 4x*BMW 003A-1* turbojet engines with 2x*HWK 509A* bi-fuel powered rocket engines. This version is known as the *Ar 234C-3R.1a*. Its *HWK* rocket engines, housed in streamlined nacelles, would occupy the same underwing positions as did the *BMW 003s*. This engine arrangement would have looked similar to the proposed 2x*HWK 509A*-equipped *Me 262 Interzeptor III*. *Van Nes* abandoned the *234C-3R.1a* idea and, instead, wanted to take a single *HWK 509C*, placing the power unit inside the tail of the *C-3* similar to the *DFS 228*. This was known within *Arado* as the *Ar 234C-3R.1b*. *Hellmuth Walter* had originally designed his *HWK 509C*, with its dual combustion nozzle, for the *Me 263* rocket-powered fighter/interceptor machine. The *509C's* upper or main nozzle produced 3,307 [1,500 kilograms] pounds thrust. The lower chamber known as the auxiliary cruising chamber provided 882 [400 kilograms] pounds of thrust. A rocket-powered *C-3R.1b* version would take off by via the thrust of its tail-mounted rocket engine. Afterward, it would climb up

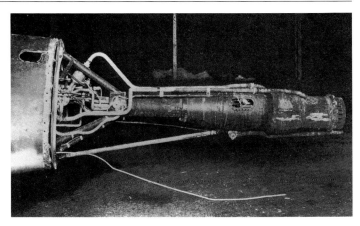

The *HWK 509A* bi-fuel liquid rocket engine...thrust tube, combustion chamber, and nozzle as installed in the rear fuselage of an *Me 262*.

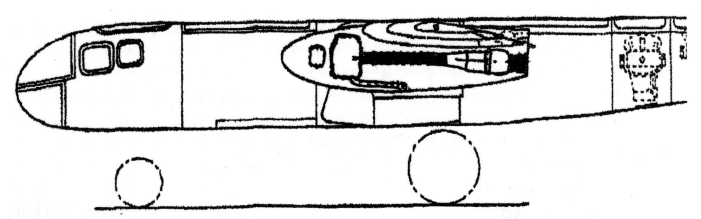

A pen and ink drawing featuring the port side of an *Ar 234C-3R* with an *HWK 509A* bi-fuel liquid rocket engine installed in place of its turbojet engine. This arrangement was never tried.

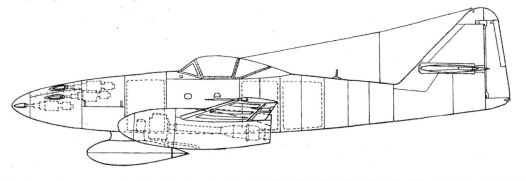

A pen and ink drawing of an *Me 262* with an *HWK 509A* bi-fuel liquid rocket engine installed in place of its normal *Jumo 004B* turbojet engine. This arrangement was never attempted.

to 54,138 [16,500 meters] feet and begin its photo reconnaissance with the knowledge that no Allied fighter could touch it at this altitude. The *509C's* auxiliary chamber would help bring the *C-3R.1b* back home.

Van Nes proposed a second version of his rocket-powered *C-3R*...with an even more powerful *HWK* bi-fuel liquid rocket engine known within *Arado* as the *Ar 234C-3R.2*. *Van Nes'* idea with this version would be propelled by a *HWK* bi-fuel rocket engine producing 4,410 [2,000 kilograms] pounds of thrust. This machine would not take off under its own power but be taken aloft piggy-back style atop a *Heinkel He 177 "Greif"* up to 26,246 [8,000 meters] feet and released. Upon release, the *C-3R.2* would climb to 55,775 [17,000 meters] feet...the same means used to launch *DFS' 228*. The *DFS 228* was being carried aloft by a *Do 217*. However, development work on a *HWK* bi-fuel rocket powered *Ar 234C-3R* design was abandoned at *Arado* when the *RLM* chose the *HWK*-powered *DFS 228*. This machine was already being test-flown although it had not been test flown with its *HWK 509C* bi-fuel liquid rocket engine operating. Nevertheless, *DFS' 228* was expected to reach 75,500 [23,000 meters] feet altitude...a height which a bi-fuel liquid rocket powered *234C-3R* simply couldn't match.

Center: A pen and ink drawing from *Arado* featuring their *Ar 234C-3R* powered only by *2x HWK 509A* bi-fuel liquid rocket engines.

Bottom: An *Arado* pen and ink port side view of the *HWK 509C-1* rocket powered *Ar 234C* high altitude photo reconnaissance machine. #A - Fuselage cross section immediately aft the *HWK 509C-1* pressurizing turbine/pumps. #B - Fuselage cross section immediately aft the vertical stabilizer and *HWK 509C-1's* thrust nozzles.

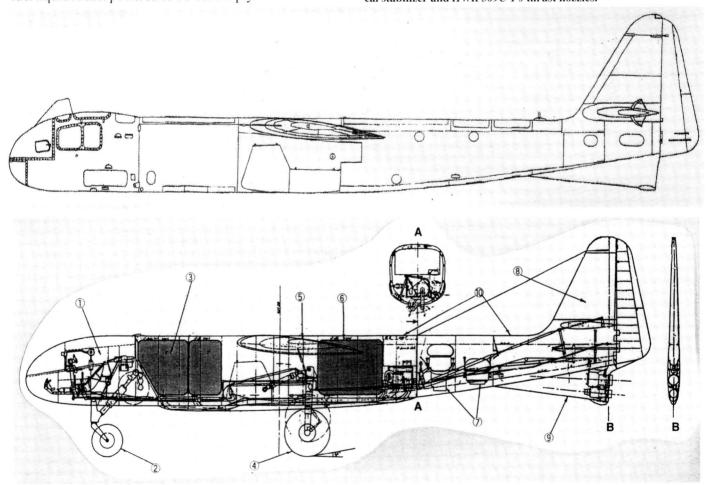

Arado Ar 234C

| Arado | Nebenverwendungszweck | Ar 234 |

d) Kurzstreckenhöhenaufklärer

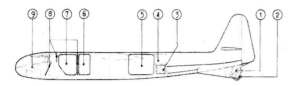

1.) Steigofen
2.) Marschofen
3.) Turbinen-Pumpenaggregat
4.) Zusatzbehälter für T-Stoff
5.) T-Stoff-Behälter (hinten
6.) T-Stoffbehälter (vorn)
7.) C-Stoffbehälter
8.) Panzerung
9.) Druckkabine

Rüstgewicht	3500 kg	
Betriebsstoff	3500 kg	(4000 kg)
Besatzung	100 kg	
Startgewicht	7100 kg	(7600 kg)
Steigschub	2000 + 400 = 2400 kg	
Marschschub	400	kg
Flächenbelastung G/F	263 ÷ 133,5	kg/m²
Schubbelastung G/Go	0,338	kg/kg
Zuladungsverhältnis B/Go	0,493	kg/kg

1.) **Reichweite:** Zur Erzielung brauchbarer Reichweiten muß das Flugzeug auf Höhe geschleppt werden. Vorgesehen ist als Schleppflugzeug die He 177 mit DB 610. Dabei sind beim Start die Flächenbelastungen beider Flugzeuge etwa gleich. Als Schlepphöhe können etwa 8000 m erreicht werden. Unter dieser Voraussetzung werden die Leistungen gerechnet.

2.) **Triebwerk:** Als Triebwerk wird das Walter-Schubgerät mit 2000 kg Steigschub und 400 kg Marschschub verwendet. Steig- und Marschofen werden unter das Rumpfheck gebaut, sodaß am Flugzeug selbst nur geringe Bauänderungen vorgenommen werden müssen. Die Leitungen von den Behältern zur Pumpenanlage und weiter zu den Öfen werden außenbords verlegt und abgedeckt.

3.) **Antriebe:** Zum Antrieb von Atemluftpresser, Hydraulikpumpe und Generator wird entweder ein Seppelerraggregat, das unter dem Flügel an dem TL-Aufhängebeschlägen angebracht ist, oder eine kleine T-Stoff-Turbine mit Getriebe verwendet.

4.) **Druckkabine:** Als Druckkabine soll eine vorgezogene Kabine der Ar 234 C angebaut werden, was einige bauliche Änderungen im vorderen Rumpfspant erfordert.

Bl.1

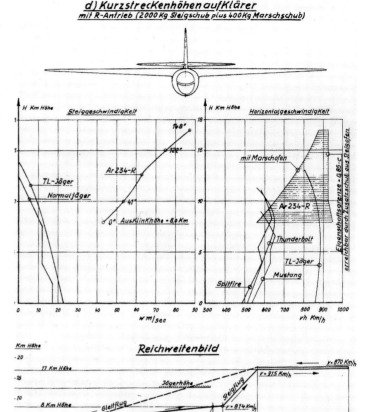

An *Arado* document dated 20 April 1944, mentions a variation of their *Ar 234C* which is powered only by a single *HWK 509C-1* dual (over and under) combustion chamber/nozzle bi-fuel liquid rocket engine. The main chamber produced 4,410 pounds [2,000 kilograms] while the so-called cruising chamber, or lower chamber, produced 660 pounds [299 kilograms] of thrust.

#01 - Upper or main chamber on the *HWK 509C-1*.
#02 - Lower or cruising chamber on the *HWK 509C-1*.
#03 - *HWK 509C-1*'s pressurizing turbine.
#04 - *C-Stoff* pressurizing pump.
#05 - Forward *T-Stoff* tank.
#06 - Aft *T-Stoff* tank.
#07 - *C-Stoff* tank.
#08 - Armor metal plating.
#09 - Cockpit cabin.

It appears from the *Arado* document dated 20 April 1944, that the *HWK 509C-1* bi-fuel liquid rocket powered *Ar 234C* would be towed up to 26,246 feet [8 kilometers] altitude (a *He 177* is mentioned as a towing machine) and then fly off to its reconnaissance target at 55,774 feet [17 kilometers] altitude powered by its main *HWK 509C-1* chamber. Perhaps the main chamber would then be turned off and the cruising chamber used off and on, to take the machine back up to altitude as necessary as it slowly drifted down.

Arado Ar 234C

A port side view of the dual combustion chambers of the *HWK 509C-1*. This bi-fuel liquid rocket engine featured a second smaller combustion chamber directly beneath the upper "main or take-off" chamber. This smaller chamber is sometimes referred to as the "cruising chamber." *Arado* claimed that their *HWK 509C-1* powered flying machine would operate comfortably at 55,774 feet [17 kilometers] altitude and reach a forward level air speed of 540 mph [870 km/h].

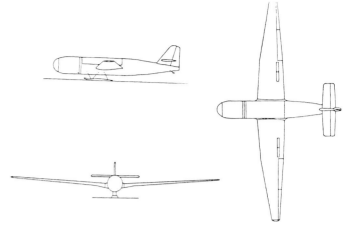

A pen and ink 3-view drawing of the *DFS 228* high altitude photo reconnaissance aircraft, powered by a single *HWK 509A* bi-fuel liquid rocket engine.

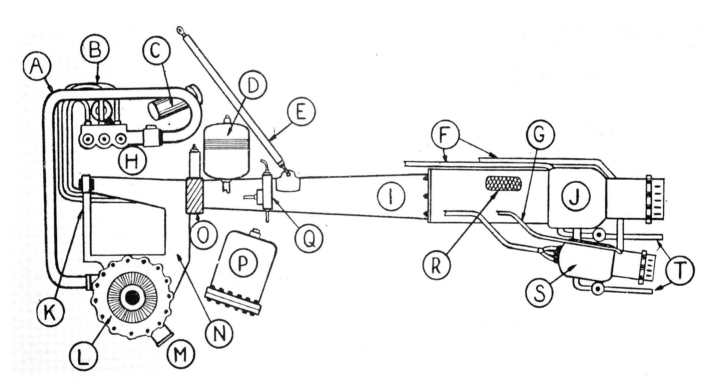

A pen and ink drawing of a complete *HWK 509C-1* bi-fuel liquid rocket engine as seen from its port side:
#A - main "*T-Stoff*" feed pipe.
#B - "*T-Stoff*" pipes to injectors.
#C - coolant filter.
#D - starting reservoir.
#E - support column.
#F - coolant pipes for main chamber.
#G - coolant pipes for auxiliary chamber.
#H - main control valves.
#I - thrust tube containing feed pipes to main chamber.
#J - main combustion chamber.
#K - thrust support tube webbing.
#L - turbine pump group.
#M - steam exhaust.
#N - turbine support.
#O - starting valve.
#P - steam generator.
#Q - ejector valve.
#R - inspection port.
#S - auxiliary combustion chamber.
#T - scavenge pipes.

Arado Ar 234C

The *DFS 228* seen high over head and appearing a lot like the *Lockheed U2*.

A *DFS 228* perched atop a *Dornier Do 217E*...its carrier aircraft throughout its testing and perhaps during operational action, as well.

The *DFS 228* high altitude photo reconnaissance machine as seen from the rear. Digital image by *Mario Merino*.

Arado Ar 234C-4

Mission: Long-Range Photo Aerial Reconnaissance

Mission-Carrying Equipment:
- 2x*Rb 75/30*, or 2x*Rb 50/30*, or 2x*Rb 20/30* cameras mounted in the aft fuselage forward the tail plane assembly;
- 2x*MG 151 20* mm[250 rounds per cannon] fixed cannon firing rearward in rear fuselage...for self-defense;
- 2x*MG 151 20* mm [250 rounds per cannon] fixed cannon firing forward in the nose beneath the cockpit...for self-defense;
- 2x66 gallon gasoline external drop tanks...one under each of the dual engine nacelles;

Status:
- Several dedicated photo reconnaissance *Ar 234C-4* flying machines are known to have been constructed prior to war's end and if they were used for photo reconnaissance. However, it is believed that a *Ar 234C-4*, which had been delivered to *1.F/123*, had been shot down on 4 April 1945 near the town of Böblingen. It is not known for sure if this *Ar 234C-4* was carrying out photo reconnaissance when it was shot down or being test-flown;

Arado Ar 234C

A pen and ink drawing from *Arado* featuring their *Ar 234C-4*, a long-range photo aerial reconnaissance machine.*

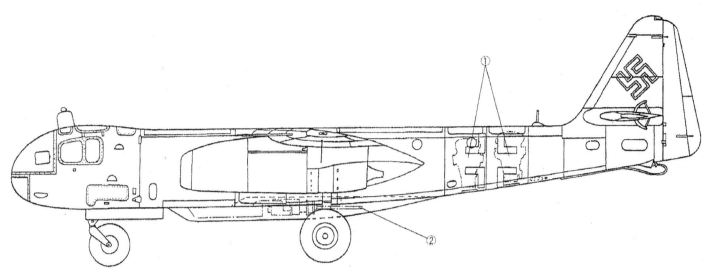

An *Arado* document from 26 September 1943, featuring a port side view of their *Ar 234C-4* reconnaissance machine and highlighting its internal photo reconnaissance camera and cannon arrangement.
#01 - 2x*MG 151* 20 mm forward-firing cannon with 250 rounds each.
#02 - 2x*MG 151* 20 mm rear-firing cannon with 250 rounds each.
#03 - rear-viewing periscope.
#04 - 2x*Rb 75/30* photo reconnaissance cameras. Other cameras could be accommodated such as 2x*Rb 50/30*, 1x*Rb 75/30* plus 1x*Rb 20/30*, or 1x*Rb 50/30* plus 1x*Rb 20/30*.
#05 - bomb rack installed to the main spar and located between the 2x*BMW 003A-1* turbojet engines.
#06 - forward and aft *J2* turbojet engine fuel tanks.

Left: An *Arado* document from 26 September 1944, featuring the general arrangement of their single seat *Ar 234C-4* reconnaissance machine. Notice that its electronic installation includes:
- *FuG 15*
- *FuG 25a - Erstling* IFF system
- *FuG 217*
- *FuG 136*
- *FuG 142*

In addition, its camera arrangement for reconnaissance duties could include one of the following:
- 2x*Rb 75/30*
- 2x*Rb 50/30*
- 1x*Rb 75/30* plus 1x*Rb 20/30*
- 1x*Rb 50/30* plus 1x*Rb 20/30*

Arado Ar 234C-5

Mission: Long Distance Conventional Bomber

Mission-Carrying Equipment:
- 2 man side-by-side cockpit;
- 1x*Schloss 500* or *2000* bomb rack underside of fuselage;
- 1x*ETC 504* rack underside each dual engine nacelle;
- Periscope with internal *BZA 1-B* bombing computer;
- *Lotfe 7K* bombsight;
- 2x*MG 151 20* mm [250 rounds per cannon] fixed cannon firing rearward in rear fuselage...for self-defense;
- 2x*MG 151 20* mm [300 rounds per cannon] fixed cannon firing forward in the nose beneath the cockpit...for self-defense;
- 2x66 gallon gasoline external drop tanks...one under each of the dual engine nacelles;
- 4,410 pound [2,000 kilogram] bomb load with combinations including free-falling, guided, and glide bombs:
    - SC 250
    - SB 1000 (with ring)
    - PC 1400
    - BT 400
    - SC 500
    - SC 1000
    - PC 1600
    - BT 700
    - SC 500RS
    - SD 1000
    - BT 200
    - BT 1400
    - SD 500

Status:
- No Ar 234C-5 bomber versions from the basic *Ar 234C-3* are known to have been built prior to war's end although the *RLM* had ordered as many as 3,000 *Ar 234C* bombers.

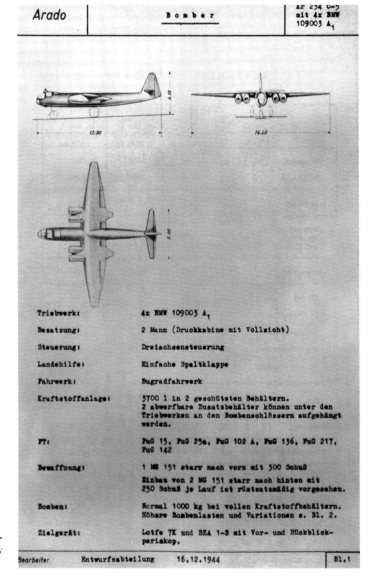

Right: An *Arado* company document from 16 December 1944, featuring the two seat *Ar 234C-5* "bomber" prototype, and powered by 4x*BMW 003A-1* turbojet engines.

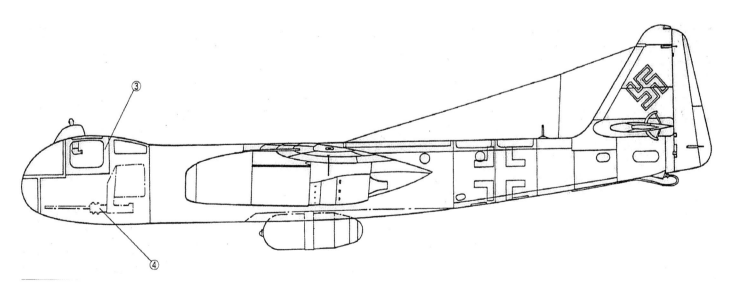

A pen and ink drawing from *Arado* featuring their *Ar 234C-5*, a long distance conventional bomber.*

Arado Ar 234C

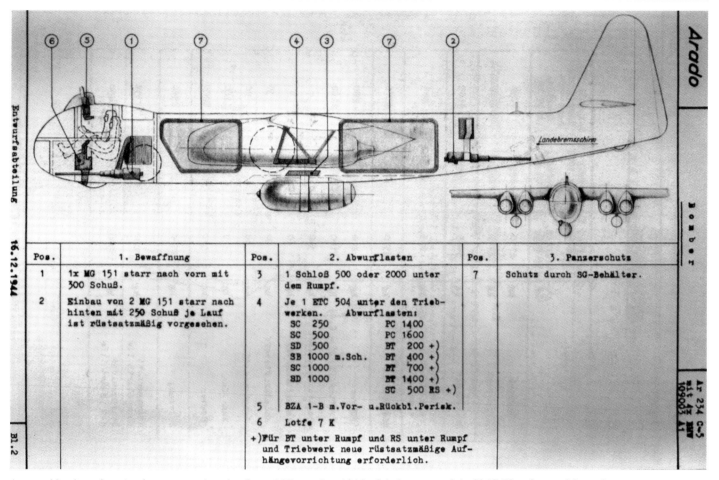

A port side view of an *Arado* company drawing from 16 December 1944, of their proposed Ar 234C-5 bomber and featuring:
#01 - 1x*MG 151* 20 mm cannon fixed forward with 300 rounds.
#02 - 2x*MG 151* 20 mm canon fixed rear-ward with 250 rounds each.
#03 - a single auxiliary gasoline tank holding between 500 to 2,000 liters of *J2* turbojet engine fuel.
#04 - 1x*ETC 504* bomb rack fixed beneath each engine pod/nacelle.

The following bombs and missiles could be attached to the *ETC 504* bomb rack and include:
• SC 250 • SC 1000 • BT 200+ • SC 500 RS+
• SC 500 • SD 1000 • BT 400+
• SD 500 • PC 1400 • BT 700+
• SB 1000 with ring • PC 1600 • BT 1400+
#05 - 1x *BZA 1-B* periscope

The wooden cockpit cabin mockup for the proposed *Ar 234C-5* high-speed bomber. Shown is the C-5 mockup's starboard side. A prototype of this machine was constructed and was known as the *V28*. It was to have been powered by 4x*BMW 003A-1* engines producing 1,760 pounds thrust each at sea level. Its first flight was to have been on 15 April 1945, and its dispostition at war's end is unknown.

The port side view of the two man cockpit wooden mockup for the *Ar 234C-5*, which was known as the *V28*. This was the first two-man cockpit of the *Ar 234C* series...pilot and bombardier siting in staggered seats. The scheduled maiden flight of the *V28* was 15 April 1945. It is not known whether its maiden flight was accomplished at this very late date in the war, especially with Germany's surrender only weeks away.

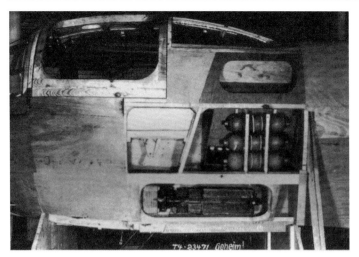

Port side wooden mockup close-up view of the two seat *Ar 234C-5N* night fighter. Two interesting features include the compartment for its three stacked oxygen cylinders and beneath it, the compartment for its *MG 151 20* mm cannon.

A view of the *Ar 234C-5N* night fighter's cockpit cabin. Featured is the pilot's seat and control column which appears in the left-hand side of the photo.

Interior, port side details, of the wooden cockpit cabin mockup of *Ar 234C-5N*. At this stage, the cockpit cabin mockup is a fairly crude example when compared to the other mockups built by the *Arado* engineers. But *Arado* engineers continue to improve on it as later photographs will show.

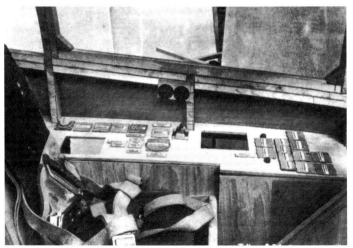

Interior port side cockpit cabin mockup of the *Ar 234C-5*, a proposed night fighter version.

Arado Ar 234C

A view from outside the advanced wooden cockpit cabin mockup of the Ar 234C-5N night fighter showing the pilot's control column and its port side windows.

A wide view of the wooden cockpit cabin mockup of the proposed Arado Ar 234C-5. Notice that this 234C-5 version has only a single pilot seat.

Another view of the cockpit cabin mockup featuring the starboard side of the proposed *Ar 234C-5*.

The proposed *Ar 234C-5* wooden cockpit cabin mockup featuring the rear bulkhead in the cockpit. The control column is barely visible to the far right of the photograph.

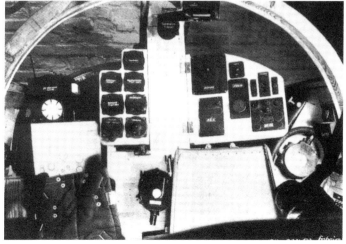

A view of the proposed *Ar 234C-5's* instrumentation with its two cockpit seats positioned side-by-side.

Arado Ar 234C

A view of a proposed *Ar 234C-5* version with two side-by-side seating rear cockpit cabin showing the seat backs and seat belts and harness.

The proposed *Ar 234C-5* featuring the items behind the navigator/bombardier's cockpit cabin seat. This view is shown from outside the starboard side of the machine.

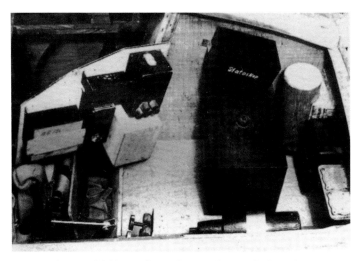

A view of the *Ar 234C-5's* aft wooden mockup cockpit cabin.

A view of the *Ar 234C-5* wooden cockpit cabin mockup as seen from the port side window looking in. Visible is the machine's periscope (in the center of the photograph) inclined forward in the retracted position.

Ar 234C-5 with *Hs 294D* Torpedo Bomb

One variation of the *Ar 234-C-5* was to have been fitted with a *Henschel Hs 294* glider torpedo bomb and a *KG* bombsight. This *234C-5* flying machine was known as "*Rüstsatz*" modification "*T1.*" The *Hs 294* had a underwater shape resembling a conventional torpedo. Once released from the carrier-aircraft, it was guided so that it entered the water at a distance of about 100 feet [30 meters] from the target naval ship. Upon entering the water its short wings and tail fins tore away immediately. As the torpedo bomb approached the target ship driven by its own momentum, it could be guided and detonated under the ship. Plus, it also was fitted with a percussion fuse. The *Hs 294* torpedo bomb weighed about 4,409 pounds [2,000 kilograms] and testing had not been completed at war's end.

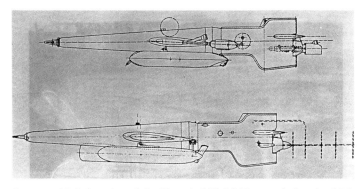

A pen and ink drawing of the *Henschel Hs 294D* torpedo bomb which was to be carried by the proposed *Ar 234C-5*. The upper view is the internal items of the *Hs 294D* including its *HWK* bi-fuel liquid rocket engine. The lower view features its trailing antenna.

Arado Ar 234C

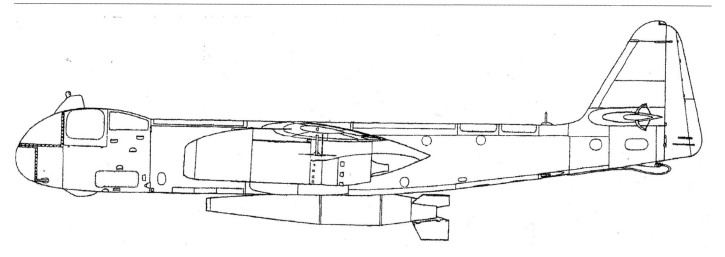

A pen and ink drawing from *Arado* featuring the port side of their *Ar 234C-5*, *Hs 294D*- carrying torpedo bomb.*

Ar 234C-5 with *Hs 295* Air-To-Ship Armor-Piercing Missile

One variation of the *Ar 234-C-5* was to have been fitted with a *Henschel Hs 295* guided air-to-water armor-piercing missile with capability to attack surface ships with its 2,205 pound [1,000 kilogram] explosive charge. This missile would have used the *FuG 203* system for guidance. The *Hs 295 "Rüstsatz"* modification for the *Ar 234C-5* was known as *"F1"* and this version of the *Hs 295* had been specifically designed to be carried under the aircraft's fuselage on the four-engined *Arado*. The *Hs 295* differed from the older *Henschel Hs 293* air-to-ground/water version in that the *Hs 295's* twin *HWK 507* bi-fuel liquid rocket engines of a combined thrust of 6,000 pounds, were placed closer to the warhead. The purpose for this was to allow for more ground clearance when hung under the *Ar 234C-5*. The *Hs 295* missile could reach about 475 mph and weighed about 4,409 pounds [2,000 kilograms]. Testing had not been completed at war's end but *Henschel*-company documents state that as many as 50 *Hs 295s* had been constructed...waiting for *Ar 234C-5s* which never came.

Below: A pen and ink drawing from *Arado* featuring the port side of their *Ar 234C-5*, *Hs 295* air-to-ship armor piercing missile. The *Hs 295* was designed especially to be carried by the *Ar 234C-5*. One change was to place its twin fuel tanks up right against the warhead/fuselage.*

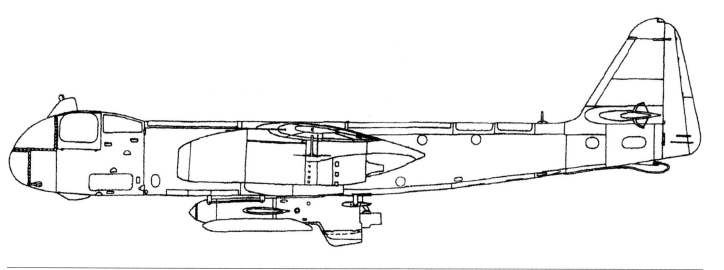

Ar 234C-5 with Bv L-11 Gliding Torpedo

The *Ar 234C-5* was to be a launch platform for the 2,205 pound [1,000 kilogram] *Blohm und Voss Bv L-11* gliding torpedo which was code-named "*Schneewittchen*" or snow white. The *Bv L-11's* "*Rüstsatz*" modification kit for the *Ar 234C-5* was known as "*T2.*" To begin with the *Bv L-11* was a standard 1,653 pound [750 kilogram] German navy fleet torpedo but with wings, tail surfaces, steering apparatus, and depth controls added. The wing and tail surface allowed the *L-11* to make a long glide before entering the water at the proper speed and angle. Upon entering the water the *L-11's* wings and tail assembly tore away leaving only the streamlined torpedo to continue on to its naval target. Approximately 450 *L-10s*, an earlier model, had been constructed. Production stopped in 1944 in order to switch over to the more accurate *L-11*. Only a limited production run had started prior to war's end and none are believed to have been tested with a *Ar 234C-3*, let alone, a *C-5*.

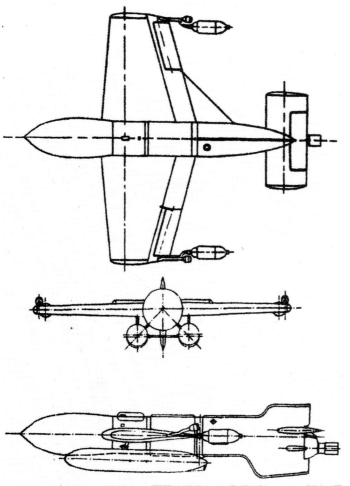

A pen and ink drawing from *Henschel* of their *Hs 295* air-to-air anti-aircraft guided missile.

A port side view of the *Hs 295* air-to-air anti-aircraft missile to be carried by an *Ar 234C-5*.

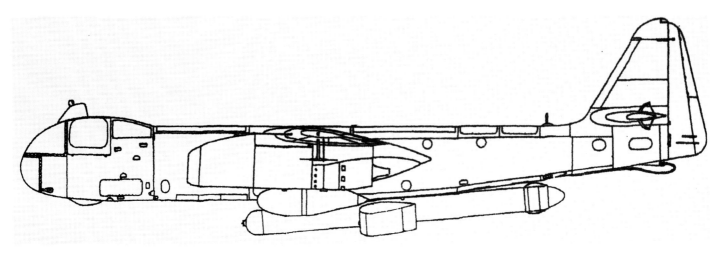

A pen and ink drawing from *Arado* of their proposed *Ar 234C-5* carrying a single *Blohm und Voss Bv L 11* gliding torpedo.

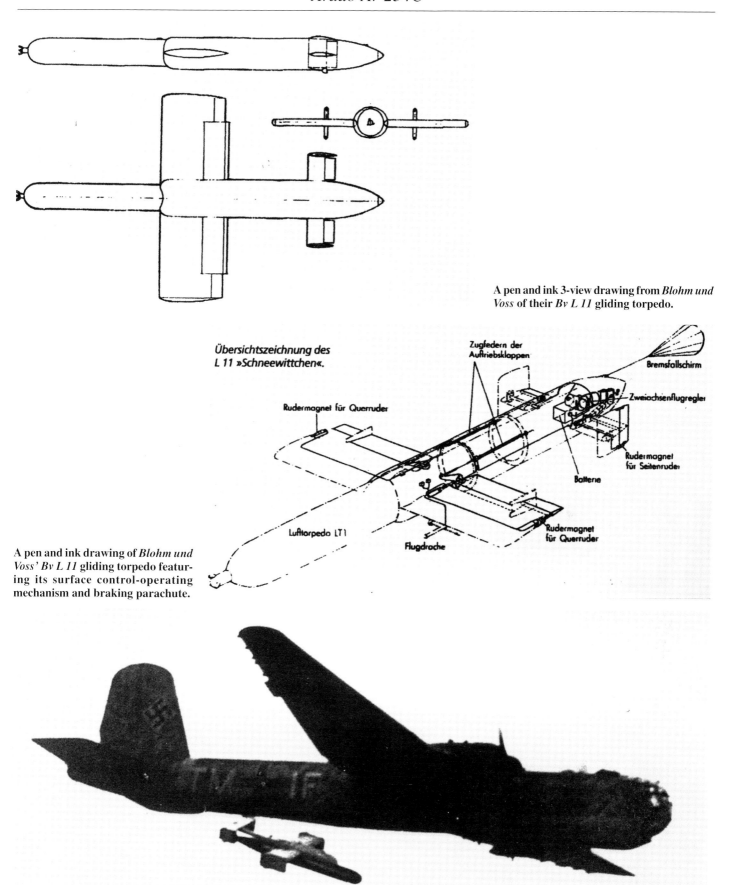

A pen and ink 3-view drawing from *Blohm und Voss* of their *Bv L 11* gliding torpedo.

A pen and ink drawing of *Blohm und Voss' Bv L 11* gliding torpedo featuring its surface control-operating mechanism and braking parachute.

A poor quality photo of a *Heinkel He 177* moments after releasing its *Blohm und Voss Bv L 11* gliding torpedo during test activities.

Arado Ar 234C

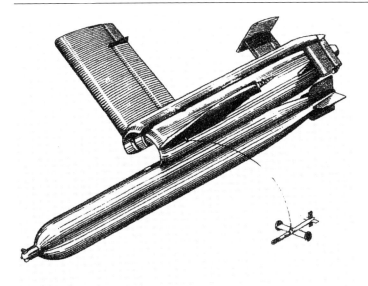

A pen and ink drawing from *Blohm und Voss* of their *Bv L 11* gliding torpedo shedding its wings and tail assembly immediately upon hitting the water. As it entered the water, the *L 11* became as aerodynamically smooth as a regular naval torpedo released from a submarine or surface vessel.

Ar 234C-5 with 2x*Daimler-Benz 021* Turboprops

It had been anticipated that the *Ar 234C* would be able to use any of the several turbojet and turboprop engines under development in war-time Germany. One such engine was the *Daimler-Benz* (*DB*) *021* turboprop rated at 3,300 shaft horsepower and 1,742 [790 kilogram] thrust at a speed of 560 mph. This turboprop turned a 5½ [2.5 meter] foot six-bladed propeller. With the turboprop, *Arado* engineers believed that their *Ar 234C-5* would benefit by experiencing quicker take-offs and more safe landings, especially during the critical approach-making time. The *DB 021* propeller's pitch could be changed upon touch down and considerably slow the machine without the use of a braking parachute. The *DB 021* was not a *Daimler-Benz* original design and they really had nothing to do with its development. It had been designed and developed by *Heinkel-Hirth* as their *HeS 011A* and adapted for driving a propeller. As a turboprop, it was designed as the *HeS 021*. The *RLM* assigned *Daimler-Benz* to series produce the *HeS 021* turboprop as the *DB 021*. The *HeS 011* was Germany's 2nd generation turbojet engine and technical difficulties were immense. Nevertheless, *Heinkel-Hirth* managed to pretty much perfect their 2,860 pound thrust engine near war's end but in the end, only ten ready-to-run *HeS 011As* were built. No *DB 021s* are known to have been manufactured.

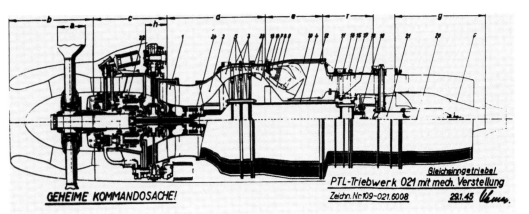

A pen and ink drawing of the proposed *DB 021* turboprop engine featuring its internal details. The *DB 021* was the *Heinkel-Hirth HeS 011A* but with a propeller fixed to its turbine shaft. It was to have been built under license to *Daimler-Benz* as instructed by the *RLM*. No working version was completed prior to war's end.

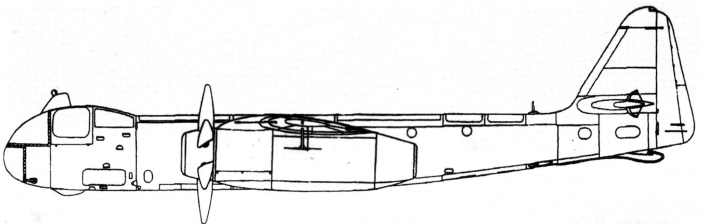

A pen and ink drawing by *Arado* of their proposed *Ar 234C-5* powered by 2x*Damiler-Benz 021* turboprops.*

Ar 234C-6

Mission: Long-Range Aerial Photo Reconnaissance

Mission-Carrying Equipment:
- Redesigned dual seat cockpit cabin similar to the *C-5* machine;
- 2x*Rb 50/30* or *Rb 75/30* standard-issue reconnaissance camera set mounted in the rear fuselage known as *Rüstsatz K1* modification kit;
- 2x*MG 151 20* mm cannon in the nose: one port side and one starboard side with 250 rounds per cannon for self-defense;
- 2x*MG 151 20* mm cannon (*Rüstsatz R2* modification kit) pod with both cannon fixed rearward with 250 rounds per cannon for self-defense;
- 2x66 gallon external fuel tanks could be attached beneath each of the two dual engine nacelles for extended range operations.

Status:

The *C-6* was to be a dual-seat photo reconnaissance aircraft. It would have used the same fuselage as the *Ar 234C*-5 series. In fact, they were identical to the *C-5*, having cameras in the rear fuselage where the *C-5*'s 2x*MG 151 20* mm cannons were located. These *C-6* aerial photo reconnaissance machines were to be powered by 4x*BMW 003A* turbojet engines. It was planned, later, to install two *Heinkel-Hirth HeS 011A* turbojet engines in place of the 4x*BMW 003A-1s*. The *Ar 234C-6* was never built nor was the *HeS 011A* turbojet engine ready for field use by war's end.

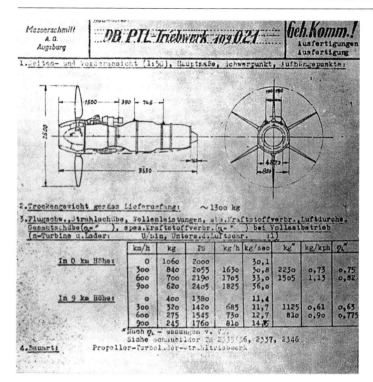

Daimler-Benz's proposed turboprop engine known as the *DB 021*, for use on the proposed *Ar 234C* variation, the *DB 021* would have been turning a six-bladed propeller.

Arado's proposed *Ar 234 PTL* and powered by 2x*DB 021* turboprop engines.

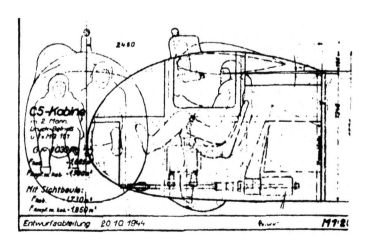

A pen and ink drawing from *Arado* featuring the cockpit cabin seating arrangement and layout for their proposed *Ar 234C-6* long range aerial photo reconnaissance machine. Notice that the navigator, although sitting side-by-side with the pilot, his seat is lower to the cockpit cabin floor.

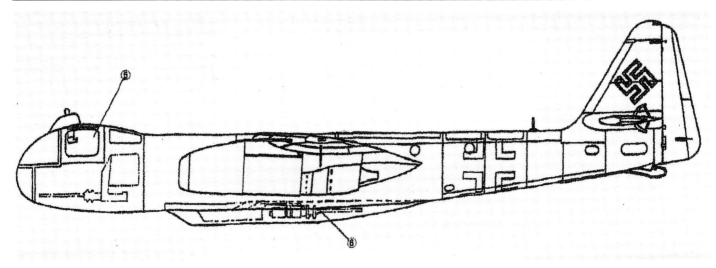

A pen and ink drawing from *Arado* featuring the port side of their proposed two seat *Ar 234C-6*, a long-range aerial photo reconnaissance machine.*

Ar 234C-7

Mission: Radar-Equipped Night-Fighter

Mission-Carrying Equipment:
- Redesigned cockpit cabin;
- *FuG* 240 *Berlin N 1a* 9 centimetric wavelength interception radar. Its parabolic dish was to have been mounted in the aircraft's extended nose and covered with a thin plywood cone;
- 2x*MG 151* 20 mm cannon in the nose: one port side and one starboard side with 250 rounds per cannon;
- 2x*MG 108* 30 mm cannon (*Rüstsatz R1* modification kit) pod with both cannon fixed forward and 100 rounds per cannon;
- 2x66 gallon external fuel tanks could be attached to extend range.

Status:
The *C-7* was to be a dual-seat nightfighter and it would have been the next to the last version of the *Ar 234C*-3 series. The fuselage was a standard *C-5* style, however, with an extended nose. These *C-7* nightfighting machines were to be powered by four *BMW 003A-1* turbojet engines. The *Ar 234C-7* was never built.

Shown is the nose-mounted parabolic scanner with rotating di-pole of the compact *FuG 240* "Berlin N-1a" radar unit. The "Berlin N-1a" of this type would have been installed in the *Ar 234C-7* and the *Ar 234 P-5*. The "Berlin N" operated on a 9 centimeter wavelength and the set was successful although it was in use for only a very short period of time. Several *Junkers Ju 88G-7c* aircraft were converted and equipped with the new radar set. A non-inductive thin plywood nose cone replaced the former array type of installation. The cleaner installation is reported to have increased the *Ju 88G-7c*'s maximum speed by at least 20 knots and is very much like today's configuration.

Arado Ar 234C

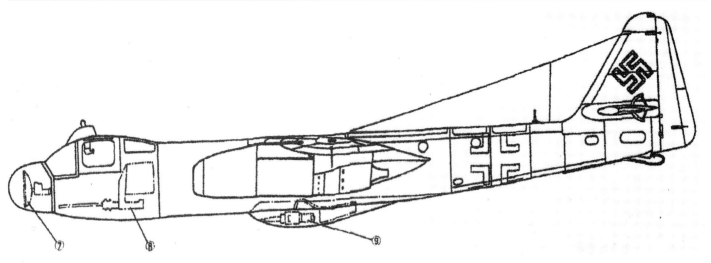

A pen and ink drawing from *Arado* featuring the port side of their proposed *Ar 234C-7* with the *FuG 240* "Berlin N" centimeteric interception radar-equipped night fighter.*

The *Ar 234C-7* would have been fitted with 2x*MK 108 30* mm cannon fixed forward in a under fuselage compartment. An *MG 103 30* mm cannon is shown here post war with *Flight Lieutenant*, and later noted aviation historian *Alfred Price*.

Ar 234C-8

Mission: Single-Seat Bomber

Mission-Carrying Equipment:
- 2x *MG 151 20* mm mounted in the fuselage: one port side and one starboard side;
- 2x*Jumo 004D* turbojet engines providing 2,315 [1,050 kilograms] pound thrust;
- Mixed external bomb load up to a maximum of 3,307 [1,500 kilograms] pounds;
- 2x66 gallon external fuel tanks could be attached to extend range.

Status:

The *C-8* was to be a single-seat bomber and it would have been the last version of the *Ar 234C* series. The fuselage was a *C-3* style, however, the *C-8* machines were to be powered by two *Jumo 004Ds*. The switch to *Jumo 004Ds*, with their 2,315 [1,050 kilogram] pounds thrust, was being considered, by *Arado* engineers, due to the continuing technical difficulties with the *BMW 003A* turbojet engine. With 2x*Jumo 004Ds*, the expected top level air speed of the *C-8* was expected to drop by 56 [90 km/h] miles per hour. Never built.

Arado Ar 234C

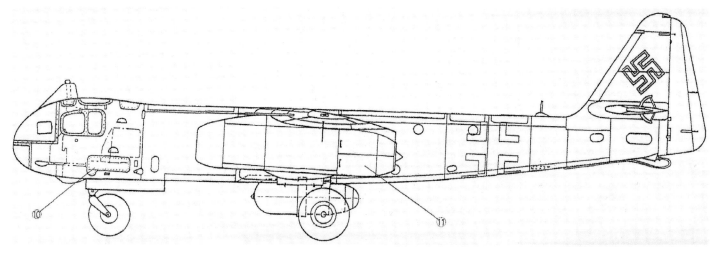

A pen and ink drawing from *Arado* featuring their *Ar 234C-8*...a proposed single-seat bomber powered by 2x*Jumo 004D* turbojet engines.*

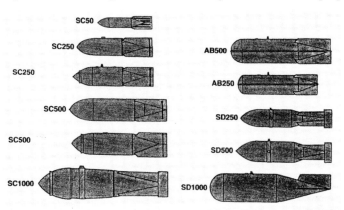

Ar 234D-1

Mission: Long-Range Aerial Photo Reconnaissance

Mission-Carrying Equipment:
- Enlarged dual seat cockpit cabin...bigger than the *C-5* machine;
- 2x*HeS 011A* turbojet engines of 2,866 [1,300 kilograms] thrust each;
- 2x*MG 151 20* mm cannon (*Rüstsatz R2* modification kit) pod with both cannon fixed rearward with 250 rounds per cannon for self-defense;
- 2x66 gallon external fuel tanks could be attached beneath each of the two *HeS 011A* engine nacelles for extended range operations.

Status:

The *D-1* was to be a dual-seat photo reconnaissance aircraft. Its fuselage would have been lengthened to 41 feet 9 inches[12.78 meters] overall. Take-off weight was approximately 21,715 [9,850 kilograms] pounds. These *D-1* aerial photo reconnaissance machines were to be powered by 2x*HeS 011A* turbojet engines. The *Ar 234D* series was never built because the *HeS 011A* turbojet engine was not ready for field use by war's end.

Above Left: *Luftwaffe* bomb choices: from the *SC 50* to the *SD 1000*...*any* of which and in combinations could be carried by the proposed *Ar 23C-8*.

Left: The complete two-man pressurized cockpit cabin for the *Ar 234D-1* and *E-1* "Destroyer" in the form of a wooden mockup.

Ar 234D-2

A view into the interior of the two man pressurized cockpit cabin for the *Ar 234D-1* and *E-1 "Destroyer"* in the form of a wooden mockup.

Mission: Long-Range Bomber

Mission-Carrying Equipment:
- Enlarged dual seat cockpit cabin...bigger than the *C-5* machine;
- 2x*HeS 011A* turbojet engines of 2,866 [1,300 kilograms] thrust each;
- 2x*MG 151 20* mm cannon in the nose (port and starboard side) with both cannon fixed forward with 250 rounds per cannon for self- defense;
- 2x66 gallon external fuel tanks could be attached beneath each of the two *HeS 011A* engine nacelles for extended range operations.
- Up to one of the *Luftwaffe's* heaviest aerial bombs: *SB 2,500*;

Status:

The *D-2* was to be a long range bomber capable of carrying some of the *Luftwaffe's* heaviest bombs such as the *SB 2500* (2,500 kilograms or 5,512 pounds). In order to carry out its mission, the *D-2*'s fuselage would have been lengthened to 41 feet 9 inches[12.78 meters] overall. Take-off weight was approximately 21,715 [9,850 kilograms] pounds. These *D-2* long-range bombers were to be powered by 2x*HeS 011A* turbojet engines. The *Ar 234D* series was never built because the *HeS 011A* turbojet engine was not ready for field use by war's end.

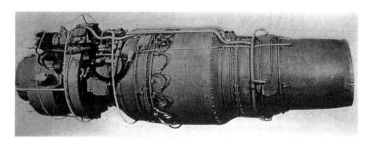

A port side view of the *Heinkel-Hirth HeS 011A* turbojet engine with a maximum thrust output of 2,680 pounds at sea level.

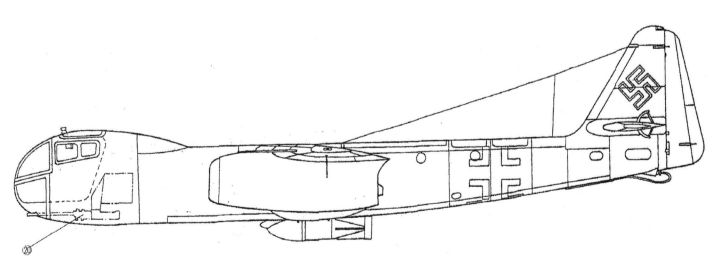

A pen and ink drawing by *Arado* featuring their proposed *Ar 234D-2* long range bomber.*

Arado Ar 234C

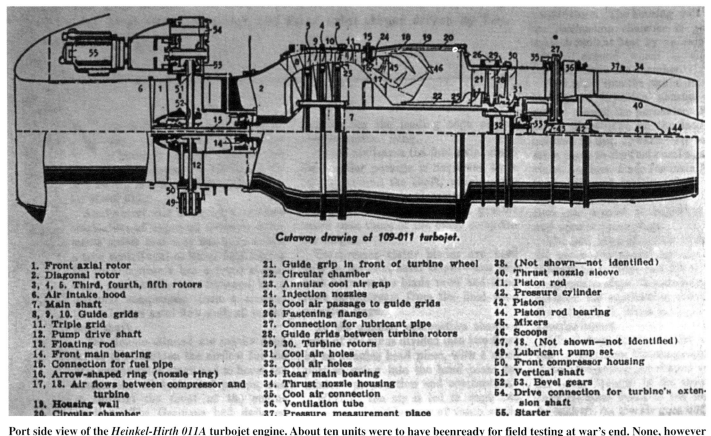

Cutaway drawing of 109-011 turbojet.

Port side view of the *Heinkel-Hirth 011A* turbojet engine. About ten units were to have been ready for field testing at war's end. None, however were.

#01 - forward axial rotor
#02 - diagonal rotor
#03 - #3 rotor
#04 - #4 rotor
#05 - #5 rotor
#06 - air intake hood
#07 - main shaft
#08 - #8 guide grid
#09 - #9 guide grid
#10 - #10 guide grid
#11 - triple grid
#12 - pump drive shaft
#13 - floating rod
#14 - front main bearing
#15 - connection for fuel line
#16 - arrow-shaped ring (nozzle ring)
#17 and #18 - air flow between compressor and turbine
#19 - Housing wall
#20 - circular chamber
#21 - guide ring in front of turbine wheel
#22 - circular chamber
#23 - annular cool air gap
#24 - injection nozzles
#25 - cool air passage to guide grids
#26 - fastening flange
#27 - connection for lubricant pipe
#28 - guide grids between turbine rotors
#29 - turbine rotors
#30 - turbine rotors
#31 - cool air ports
#32 - cool air ports
#33 - rear main bearing
#34 - thrust nozzle housing
#35 - cool air connection
#36 - ventilation tube
#37 - pressure measurement point
#38 - not identified
#39 - not identified
#40 - thrust nozzle sleave
#41 - piston rod
#42 - pressure cylinder
#43 - piston
#44 - piston rod bearing
#45 - mixers
#46 - scoops
#47 - not identified
#48 - not identified
#49 - lubricant pump set
#50 - front compressor housing
#51 - vertical shaft
#52 - bevel gears
#53 - bevel gears
#54 - drive connection for turbine's extensive shaft
#55 - starter

Ar 234D-E.395

Mission: Enlarged version of the Ar *234D-2* bomber. Known also as the Ar 234F.

Mission-Carrying Equipment:
- 4,410 [2,000 kilogram] pound bomb load;
- 4x*MG 151 20* mm tail turret, in the twin rudder version;
- 4x*MK 103* cannon in the destroyer version;
- 2x*MG 151 20* mm cannon in the destroyer version;
- 3x*Fritz-X* air-to-air guided missiles;

Status:

The *Ar 234D-E.395* was proposed by *Rüdiger Kosin's* design-development team in late 1944 as a larger bomber version of their *Ar 234D-2*. *RLM's* plan was that this new, larger machine would replace the *Heinkel He 177* "*Greif.*" The *E.395* machine would have been powered by 4x*HeS 011A* turbojet engines with 2,860 pounds thrust each. *Each* of the turbojets would have been mounted in individual nacelles, similar to the four motored *Ar 234V6*. *Kosin's* team also made provisions for the *E.395* to be powered by alternative engines such as 4x*BMW 003C*, 4x *Jumo 004C*, or 2x*Jumo 012* with their estimated design thrust of between 6,000 and 6,400 pounds. The *E.395's* maiden flight was planned for 31 December 1944. How-

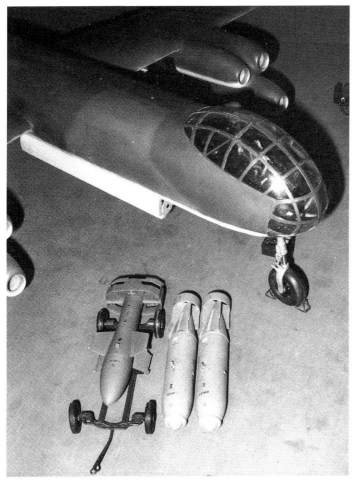

A six turbojet motored *Ju 287 V3* in the background. In the foreground is a *Fritz "X"* wire-controlled glide anti-shipping bomb (*PC 1400*) and 2x*SC2000* bombs. The "*Fritz-X*" is on the left. Both of these items were to have been carried, too, by the *Ar 234C*. Scale models and photographed by *Günter Sengfelder*.

An *SB* 2500 kilogram [5,512 pound] bomb. It measured 12 feet 1 inch [3.69 meters] in length and 2 1/2 feet [0.78 meters] in diameter.

Arado Ar 234C

ever, *Arado* was instructed by the *RLM* to drop the large bomber version of the *Ar 234D-2* in favor of concentrating all available in-house resources on an emergency production of the *Fw 190,* fighter under license. However, the *Ar 234D-E.395* continued to live on with all of its design plans turned over to *Heinkel AG* in May 1944. At *Heinkel AG*, the former *Kosin-Arado* design was redesignated as the *He 343. Heinkel AG's* chief designer, *Siegfried Günter* set about constructing three versions: a heavy bomber carrying a 4,410 pound bomb load, a photo reconnaissance version, and a destroyer version with twin vertical rudders. Construction started on a prototype in June 1944, with its maiden flight scheduled for April 1945. However, *Karl France, Heinkel AG's* Technical Director, is reported to have abandoned all work on the *He 343* in January 1945. Then, again, there are reports that two prototypes were under construction.

A pen and ink illustration from *Heinkel AG* of their proposed *He 343 (Ar234D-E.395)* heavy bomber version, the former *Arado Ar 234D-E.395.*

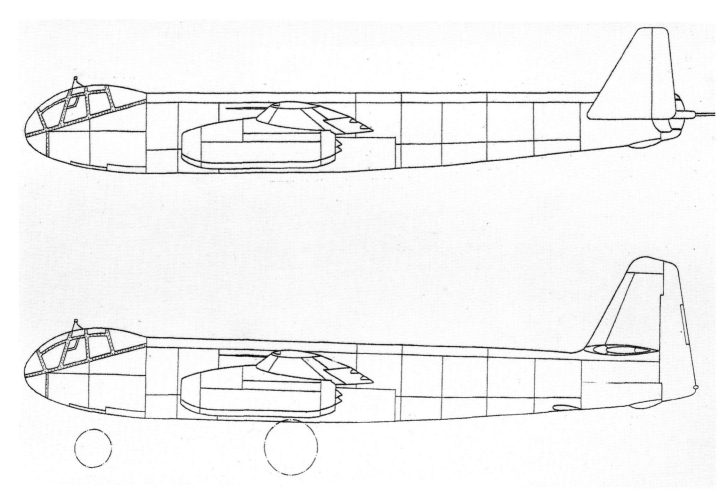

A pen and ink drawing featuring the port side of two proposed versions of the *He 343 (Ar 234D-E.395)*. Top: the dual ruddered "destroyer version." Bottom: the heavy bomber version capable of accepting a 4,410 pound bomb load.*

Arado Ar 234C

The *He 343A-1 (Ar 234D-E.395)* heavy bomber version with its single vertical rudder. Scale model and photographed by *Reinhard Roeser*.

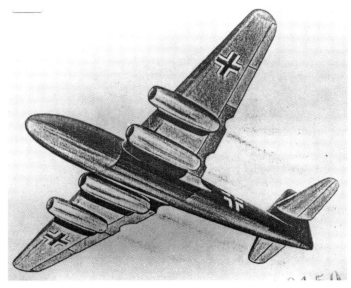

A pen and ink illustration, done by the Allies post war, of the *Heinkel AG He 343A-1 (Ar 234D-E.395)* heavy bomber version.

The *He 343 (Ar 234D-E.395)* shown here with its twin vertical rudders. This heavily armed version was known as the "destroyer" version.

The *He 343 (Ar 234D-E.395)* was also intended to have one remote control rear-firing cannon like the one shown in this photograph.

Ar 234E

Mission: Heavy Armored Ground Attack/Destroyer

Mission-Carrying Equipment:
- Enlarged dual seat cockpit cabin...bigger than the *C-5* machine;
- 2x*HeS 011A* turbojet engines of 2,866 [1,300 kilograms] thrust each or 2x*Jumo 004C* turbojet engines of 2,866 [1,300 kilograms] pounds thrust;
- 2x*MK 108* 30 mm cannon (*Rüstsatz R3* modification kit) under fuselage pod with both cannon fixed forward with 250 rounds per cannon. The *R3* kit weighed 661 [300 kilograms] pounds;
- 2x*MG 151* 20 mm cannon in the nose: one port and one starboard;
- 2x66 gallon external fuel tanks could be attached beneath each of the two *HeS 011A* engine nacelles for extended range operations.
- 2x*SC 500 RS* unguided rocket bombs;
- 3x*Magirusbombe* pods...one under the fuselage and one beneath each of the two *HeS 011A* turbojet engine nacelles;

Status:
The *234E* was to be a "destroyer" air machine capable of carrying some of the *Luftwaffe's* heaviest hitting cannon. In order to carry out its mission, the *234E*'s fuselage would have been lengthened to 41 feet 9 inches [12.78 meters] overall...similar to the *234D* series. Take-off weight was approximately 21,715 [9,850 kilograms] pounds, also similar to the *234D* series. These *234E* heavy armored attack versions were to be powered by either 2x*Jumo 004C* or 2x*HeS 011A* turbojet engines. The *Ar 234E* series was never started because neither turbojet engines were ready for field use by war's end.

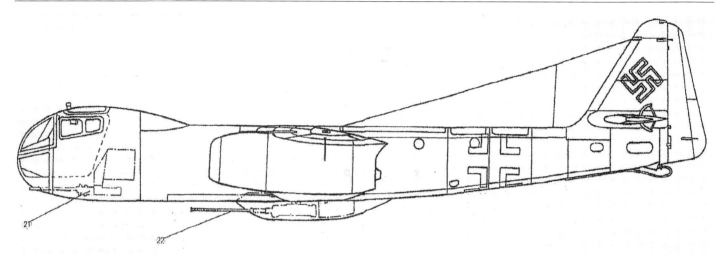

A pen and ink drawing from *Arado* featuring their proposed two-man *Ar 234E*...heavy armored ground attack/destroyer.*

The *Ar 234 D-1* would have been powered by a pair of 2,800 pound [1,300 kilogram] thrust *HeS 001A* turbojet engines. The port side of the *HeS 011A* is shown and each engine weighed 2,009 pounds. This second generation turbojet engine was not quite ready for field testing prior to Germany's surrender on 8 May 1945.

The starboard side of the *HeS 011A* turbojet engine. Its *Riedel* starter is outside the engine attached to the accessory gear housing above the air intake cowling, unlike the *BMW 003A* and *Jumo 004B* which is attached to the main turbine shaft.

Arado Ar 234P

Mission: Dedicated Night Fighters

Mission-Carrying Equipment:
- *Panzerkabine* or 13 mm armored cockpit. This is where the "P" came from as found in *Ar 234P*;
- 2x*MK 108 30* mm *Schräge Musik* oblique 70° and/or 90° upwards firing cannon directly behind the cockpit or further aft;
- 1x or 2x*MG 151 20* mm cannon in the fuselage nose beneath the cockpit;
- *FuG 244 "Bremen"* centimetric interception radar;
- 2x*ETC 504* racks for carrying external fuel tanks and located under each turbine engine nacelle;
- *Rüstsatz R-3* under fuselage containers carrying 2x*MK 108 30* mm cannon with 1,000 rounds;
- Two and three man crews;

Status:

Rüdiger Kosin and his staff at *Arado* began plans for converting their *Ar 234C* into night and all-weather fighters about January 1945. This order came down from the "night and all-weather commission" lead by *Kurt Tank* of *Focke-Wulf*. Without the modifications, the *Ar 234C* was found to be unacceptable as a night fighter. Develop-ment work continued at *Arado* right up to Germany's surrender on 8 May 1945. The *Kosin* team was plagued with shortages of components. New, powerful turbojet engines, didn't materialize as promised, and the *FuG 244 "Bremen"* 9 centimetric wavelength interception radar was in extreme short supply. Consequently, *Kosin* was left to use the older *FuG 218 "Neptun"* radar and *BMW 003A-1* turbojet engines which had to be repositioned fore or aft depending on the night-fighting equipment installed in or hung from the fuselage. No *Ar 234P* night and all-weather fighters saw action in the late hours of WWII.

A pen and ink drawing/text by *Arado* featuring their *Ar 234E* heavy armored ground attack/destroyer's 3x*Magirus-Bomben* containing two or more *MG 151 20* mm cannon.*

Ar 234P-1 Night Fighter

This was the 1st prototype night fighter by the *Kosin* design team. *Kosin* took their *Ar 234C-7* design and modified this machine with all the necessary night fighting equipment/components necessary to serve in this role. The *Ar 234P-1* prototype had a two seat, 13 mm armored cockpit. To help off set the added weight in the nose, its 4x*BMW 003A-1* engines were repositioned 1foot 3 inches [0.4 meters] aft. Armament included 2x*MG 151 20* mm cannon in the fuselage beneath the cockpit with one cannon installed to port and the other to starboard. Installed, also, was a *Rüstsatz R3* container kit beneath the fuselage housing 2x*MK 108 30* mm cannon with 1,000 rounds. One *ETC 504* rack was installed under each of the two turbojet engine nacelles to carry a external fuel tank. A *FuG 244 "Bremen"* 9 centimetric wavelength interception radar was installed in the *Ar 234P-1*'s extended nose. Radio equipment included: *FuG 15, FuG 25a, FuG 120a, FuG 130, FuG 136*, and the *FuB1 II F*. Take off weight of the former *Ar 234C-7*, now *Ar 234P-1*, ranged between 23,590 and 25,904 pounds [10,700 and 11,750 kilograms] due to all the changes and additions. Maximum speed was estimated to be 536 mph [862 km/h] with a range of 590 miles [950 kilometers].

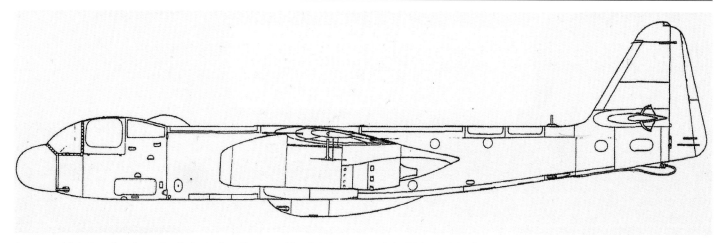

A pen and ink drawing from *Arado* featuring the port side of their proposed *Ar 234P-1* night fighter series.*

An *FuG 240 "Berlin N-1a"* radar unit on the ground. Its overall size can be judged relative to the RAF person to the right.

Ar 234P-2 Night Fighter

The 2nd prototype version of the *Ar 234P* night fighter featured a redesigned cockpit cabin. It was not a dual seat cockpit anymore...but a single seat cockpit. The radar operator, formerly seated with the pilot in the *P-1*, was now relocated to a small compartment just forward the vertical stabilizer. This change required that the 4x*BMW 003A-1* turbojet engines be repositioned, too, that is, back to their original position on the wing. Armament included 2x*MG 151 20* mm cannon located just inside the fuselage beneath the cockpit one cannon on its port and starboard sides. In addition, the machine was fitted with the *"Rüstsatz" R3 (2x MK 108 30* mm cannon) compartment kit beneath the fuselage. Several new pieces of electronics were added to the *P-2*: *FuG 140* and a *FuG 350 "Naxos."* Take off weight of the *Ar 234P-2* was 25,575 pounds [11,600 kilograms], a top forward speed of 537 mph [864 km/h], and a maximum range of 590 miles [950 kilometers].

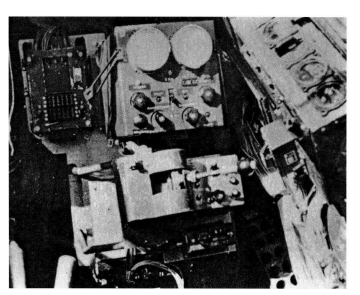

Left: The operational end of the *FuG 240* radar set. The radar operator of the airframe equipped with the *"Berlin N-1a"* set peered into the double scope. Below the double scope is the control lever of the scanner by which the radar operator could follow an enemy (Allied) bomber through all of its evasive movements trying to shake off the German nightfighter.

Arado Ar 234C

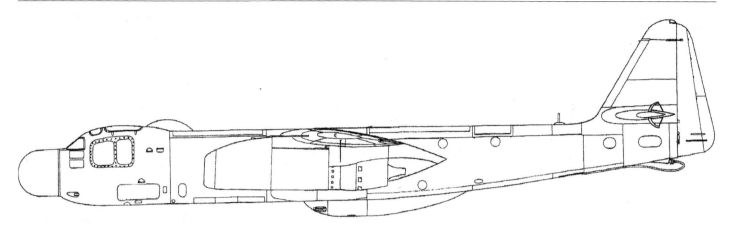

A pen and ink drawing from *Arado* featuring their proposed single seat *Ar 234P-2* night fighter.*

Ar 234P-3 Night Fighter

This 3rd *Arado* 2-person design for a *234C*-based night fighter prototype was very similar to *Kosin's* earlier *Ar 234P-3*. About the only visible difference is that the *P-3* was to have been powered by 2x*HeS 011A* turbojet engines of 2,860 pounds thrust each. With this much power, the *P-3*'s range would have increased to 882 miles [1,420 kilometers]. Take off weight was an estimated 23,534 pounds [10,675 kilograms]...less than the *P-1* or *P-2*. On the other hand, top level speed was less than the *P-1* or *P-2*, too. No explanation is given in the *Arado* documents but the reason may have been that a version powered by 4x*BMW 003A-1*s, provided a combined thrust of 6,200 pounds. The combined thrust of the 2x*Hes 011*s in the *P-3*, was 500 pounds less than 4x*BMW 003s*, with only 5,720 pounds at sea level. But the *HeS 011* could give better range but it took you longer to get there. Several *Arado* documents state that 2x*Jumo 004D* turbojet engines producing 2,200 pounds thrust each, was to be a alternative engine for the *P-3* prototype. The *Ar 234P-3*'s maximum level speed was estimated to be 510 mph [820 km/h].

Ar 234P-4 Night Fighter

The *Ar 234P-4* prototype was very similar to the *Ar 234P-3* in terms of armament and electronics, with the only exception that the *P-4* was to have been powered by 2x-*Jumo 004D's* turbojet engines of 2,200 pounds thrust each at sea level. With these turbojet engines, estimated top level speed was 450 mph [725 km/h]. Maximum range was estimated at 907 miles [1,460 kilometers]. Take off weight of the *Ar 234P-4* was 42,935 pounds [19,475 kilograms].

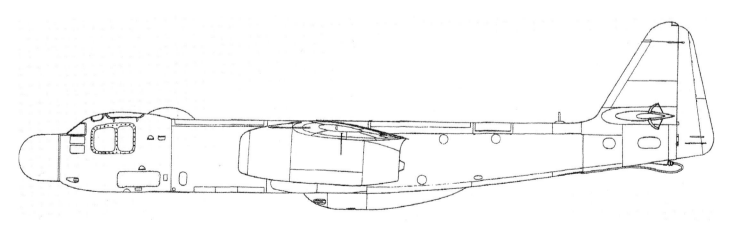

A pen and ink drawing from *Arado* featuring the port side of their proposed two man *Ar 234P-3* night fighter.*

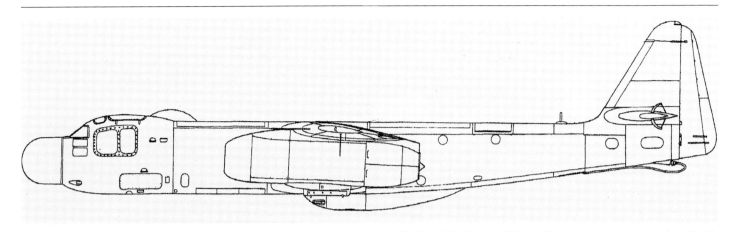

A pen and ink drawing from *Arado* featuring the port side of their two man *Ar 234P-4* night-fighter. This version was to have been equipped with 2x*Jumo 004Ds* of 2,200 pounds thrust at sea level each.*

Ar 234P-5.1 *Schräge Musik* Night Fighter

In late 1944, the *RLM* issued requirements for a *Schräge Musik* equipped night fighting machine. *Arado* submitted a proposal for a prototype *Schräge Musik* aircraft based on their *Ar 234P-2*. *Arado's* initial plans were dated 31 January 1945, and this machine is known as the *Ar 234P-5.1*. *Arado's* proposal outlined a *Schräge Musik* firing machine with a fuselage nose-mounted *FuG 244* 9 centimetric wavelength radar. When *Arado* began making plans to construct their three person *P-5.1* prototype, the *FuG 244* "*Bremen*" was unavailable so *Arado* substituted the *FuG 218* "*Neptun*" radar. The 3rd crewman had to be seated in the rear fuselage in the forward *Rb* photo reconnaissance camera bay (the same also the *P-3* and *P-4)* forward the vertical stabilizer. The second *Rb* camera bay was paneled off. *Arado* submitted a amended version of their *P-5.1 Schräge Musik* to the *RLM* on 10 February 1945, and this machine is known as the *234P-5.2*. However, their *P-5.1* was to have been powered by 4x*BMW 003A-1* turbojet engines. Its *Schräge Musik* 2x*MK108 30* mm cannon, formerly positioned in a *Rüstsatz R3* under fuselage container pod kit with fixed forward cannon and 100 rounds per cannon, were repositioned behind the cockpit cabin and mounted so to fire upwards (obliquely) at a 70° degrees. In place of the *R3* kit, *Arado* hung a 600 liter fuel tank. Additional fuel tanks were to be hung on *ETC 504* racks beneath both twin turbojet engine nacelles. In addition, the *P-5.1* carried 1x*MG 151* 20 mm nose-mounted (port side) cannon with 300 rounds. The *Ar 234P-5.1* was not built because the *RLM* chose *Arado's* amended *P-5* version, their *234P-5.2*, dated 10 February 1945. *Kosin* and his engineers hoped to have a prototype *P-5.2* ready for its maiden flight by June 1945. It is not known how far along *Arado* got with their *P-5.2* prior to war's end.

This is the 3rd crewman's (radar operator) position in the *Ar 234P-5*...where the *Rb* aerial photo reconnaissance cameras were to be located.

Arado Ar 234C

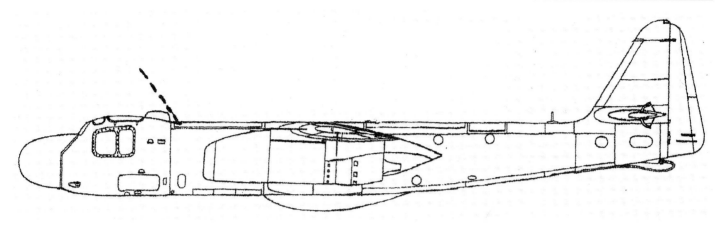

A pen and ink drawing from *Arado* featuring the port side of their proposed Ar *234P-5* night-fighter with a *FuG 244* nose-mounted *"Bremen"* radar.*

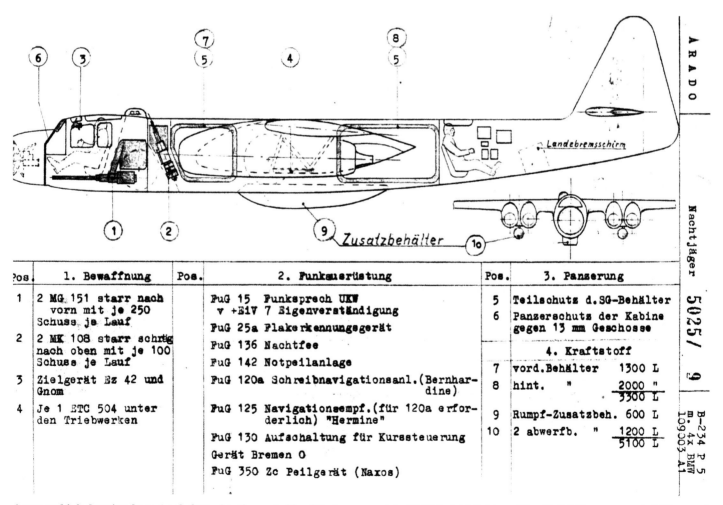

A pen and ink drawing from *Arado* featuring the port side of their proposed Ar *234P-5* night-fighter with a *FuG 244* nose-mounted *"Bremen"* radar.*

Arado Ar 234C

A Bf 110 *"Schräge Musik"* night-fighter with its three upward-firing cannon in position. A similar arrangement was planned for Arado's Ar 234P-5.1 night fighter.

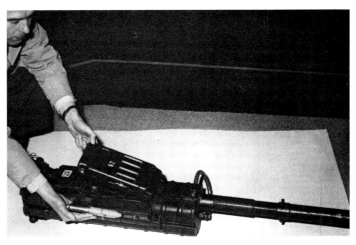

The *MK 108* 30 mm cannon of the type to be installed in the *Arado 234C-7* night-fighter. Notice the size of an *MK 108* cannon shell.

The nose, port side view of an *MK 108* 30 mm cannon of the type to have been used in the *Ar 234P-5.1 "Schräge Musik"* night-fighter.

Ar 234P-5.2 Schräge Musik Night Fighter

A prototype of the three man *Ar 234P-5.2* machine was under construction by *Rüdiger Kosin* and his team at the time of war's end. The *P-5.2's* maiden flight was scheduled for June 1945. It was to have been powered by 2x*HeS 011As*. These 2,860 pound thrust 2nd generation turbojet engines had not even entered a pre-series production at war's end, however, about ten engines available for field testing. No information exists which indicates that they were placed into any airframe...with the possible exception of the *Me P.1101* and this pure research machine had not flown by war's end either. In the absence of *HeS 011As* the *234P-5.2* perhaps would have been powered by 2x*Jumo 004Ds*.

The design to build a prototype of the *Ar 234P.5.2* came on 10 February 1945. Its cockpit and fuselage were based on the *234P-1*. Two members of the crew sat side-by-side in the cockpit. A radar operator was placed in the rear of the fuselage...sitting in the forward *Rb* reconnaissance camera bay now modified into the radar operator's quarters. Armament included 2x*MG 151* 20 mm cannon each side of the nose, a *"Rüstsatz"* R3 kit with 2x*MG 108* 30 mm with 1,000 rounds, and 2x*MG 108* 30 mm *Schräge Musik* firing obliquely upward at 70 degrees. These cannons were positioned aft of the main spar.

Several variations of the *Ar 234C* would have had 2x*MG 151* 20 mm cannon installed vertically *"Schräge Musik"* in the fuselage...shooting upward and shown in this photograph of an *Me 110*.

Arado Ar 234C

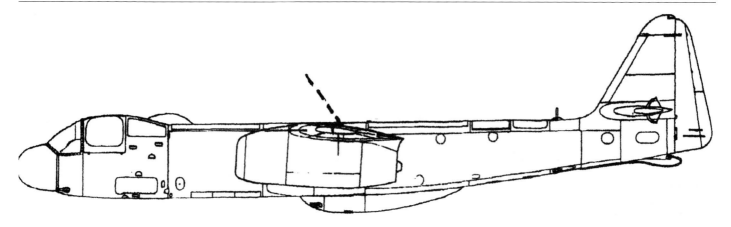

A pen and ink drawing from *Arado* featuring a port side profile of their proposed 3-man *Ar 234P-5.2* "Schräge Musik" night-fighter.*

Ar 234C-E.560

Mission: Long-Range Heavy Bomber

Mission-Carrying Equipment:
- 4,410 [2,000 kilogram] bomb load carried under the fuselage in 2x*Schloss 1000* bomb racks
- 4x*Jumo 004C* turbojets providing 2,200 [998 kilograms] pounds thrust at sea level each;
- 5x*MG 151 20* mm cannon: 1x*R1* and 1x*R2* "Rüstsatz" under fuselage pod modification kit each with 200 rounds per cannon, and one remote control cannon in the rear of the fuselage with 400 rounds;

Status:
This late war *Rüdiger Kosin* paper only design, was a proposed two man long-range heavy bomber similar in style to the *Ar 234C*. Although similar, it was much larger with a 62½ [19.10 meter] foot overall length, 59½ [18.20 meter] foot 35° sweptback wing span, 16½ [5 meter] foot overall height, and a 19½ [6.00 meter] foot wide sweptback horizontal stabilizer span. The *Ar 234C E.560*'s pilot and bombardier would have sat side-by-side in a pressurized cockpit surrounded by thick armored glass. In addition, the crew and fuel tanks would have had 5 mm to 15 mm thick armored (steel plate) protection. Radio equipment would have included the *FuG X-P*, *FuG 16*, *FuG 101*, *FuG 25a*, and the *FuB1-2*. A *K-12* auto pilot was planned. The largest *Ar 234C*-based machine ever proposed by *Arado* during the war years was, in the end, ignored by the *RLM* perhaps due to a lack of time, labor, materials, and fuel to fly it.

Rüdiger Kosin...Arado's chief aerodynamitist and head of the engineering team to convert the twin engined Ar 234B into the four engined Ar 234C machine.

Arado Ar 234C

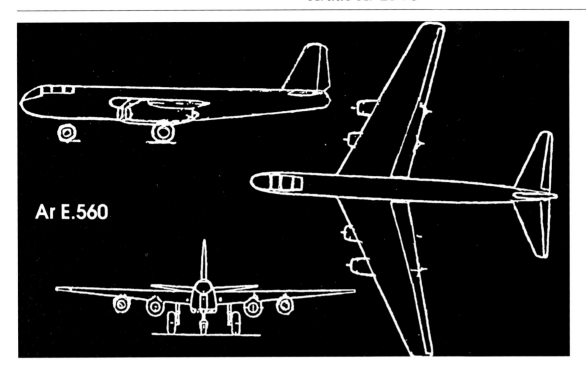

Opposite: A 3-view pen and ink drawing from *Arado* featuring their proposed two person *Ar E.560* heavy bomber. This machine was intended to carry 4,410 pounds [2x1000 kilogram] of bombs.*

A pen and ink reverse negative by *Arado* featuring a 3-view general arrangement drawing of their proposed *Ar E.560* heavy bomber.

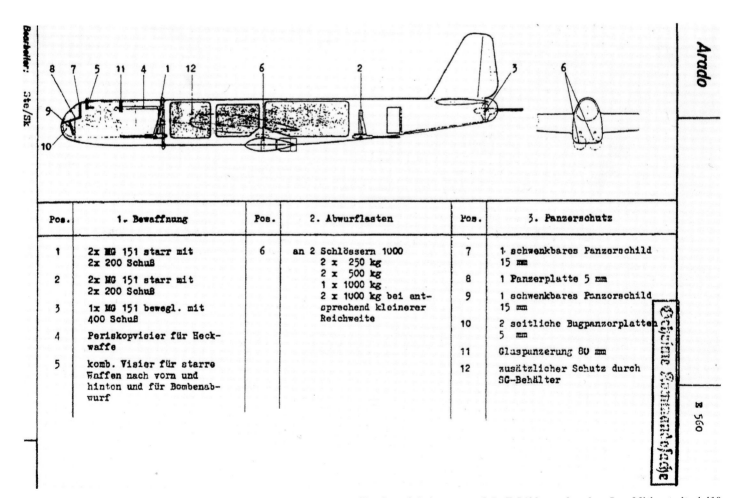

A pen and ink drawing from *Arado* featuring a port side internal profile view of their proposed *Ar E.560* heavy bomber. In addition to its 4,410 pound bomb load, the *Ar E.560* was to be heavily armed with 5x*MG 151 20* cannon: 2x*MG 151* fixed forward, 2x*MG 151* fixed rearward under the fuselage, and 1x*MG 151* in a moveable tail turret.

Arado Ar 234C

| Arado | — | E 560 |

Geheime Kommandosache

Besatzung:	2 Mann
Ausrüstung:	Druckkabine mit Vollsicht Kurssteuerung K 12 2 Einmann-Schlauchboote Kutonase
FT-Anlage:	FuG X-P, FuG 16, FuG 101, FuG 25a, FuBl 2.
Bewaffnung:	2x MG 151 starr nach vorn, 2x 200 Schuß 2x MG 151 starr nach hinten, 2x 200 Schuß 1x MG 151 i.Rumpfheck über Periskopvisier ferngesteuert, 200 Schuß 2x Schloß 1000 mit komb. Bombenabwurfzielgerät und Rückblick- visier mit Vorhalterechner
Panzerung:	Panzerung der Kabine zusätzlicher Schutz durch SG-Behälter

Arado Ar 234C

This *Arado* document dated 28 July 1944, lists month by month construction of two versions of the *Ar 234C* up through February 1946. It states that a combined total of 3,890 *Ar 234C-3* bombers and *Ar 234C-4s* reconnaissance machines will have been built by February 1946. Specifically, this production number includes 3,660 *Ar 234C-3s* and 230 *Ar 234C-4s*.

Rüdiger Kosin's final war time accomplishment...the four turbojet version *Ar 234C*. Scale model and photographed by *Günter Sengfelder*.

One of several *Ar 234Bs* used as landfill during the runway extension at Patuxent River Naval Air Station, Patuxent River, Maryland. Courtesy *J. Richard Smith* and *Eddy Creek*, *Jet Planes of the Third Reich*.

A bomb-damaged hangar at Manching, Bavaria. It was captured by the U.S. 3rd Army in April 1945. At the center of the photo is a *Ar 234B*. In addition to the *Ar 234B*, in the far right upper corner of the photograph is a *Bücker 181D "Bestman"* trainer. At the top center are 2x*Ju 88C* night fighters. Finally, at the far left corner, showing only a rudder, is actually a double ruddered *Ju 88G*.

An *Ar 234B* at war's end...destroyed by its own former ground crew.

An *Ar 234B* at war's end...destroyed by its own former ground crew.

An *Ar 234B* photographed in Germany seen virtually broken in two.

Arado Ar 234C

Walter Blume's masterpiece. The last known surviving *Ar 234* in the world is located at the *NASM*, Washington, D.C. It is an *Ar 234B-2* and it has been beautifully restored to like-new condition and is on public display.

NASM's Ar 234B-2 seen here in the United States in the mid to late 1940s when it was still being flight tested.

NASM's Ar 234B-2 as seen from its rear, starboard side in October 1964.

Dipl.-Ing. Walter Blume...1896-1964.

Notes

Notes

Notes